THE
Feynman
Processor

FRONTIERS OF SCIENCE
Series editor: Paul Davies

QUANTUM ENTANGLEMENT AND
THE COMPUTING REVOLUTION

THE
Feynman
Processor

Gerard J. Milburn
FOREWORD BY PAUL DAVIES

HELIX BOOKS

BASIC
B
BOOKS

A Member of the Perseus Books Group
New York

Library of Congress Catalog Card Number: 99-65719

ISBN-10: 0-7382-0173-1   ISBN-13: 978-0-7382-0173-3

Published in Australia by Allen & Unwin Pty Ltd

Basic Books is a Member of the Perseus Books Group.

Cover design by Suzanne Heiser
Set in 10.5-point Plantin by DOCUPRO, Sydney

Find Basic Books on the World Wide Web at
http://www.basicbooks.com

FOREWORD

It is a truism that the computer has become an icon of our age. So dazzling have been the successes of such machines, so awesome their capabilities, that there may seem to be no limit to their power and scope. Everyone knows that each year computers become faster, smaller and cheaper, and it is tempting to suppose that this trend will continue for ever.

A sober analysis reveals a more complicated picture. However magical they may appear, computers are not ideal abstract entities: they are made of material bits and pieces. As such, they are subject to the same laws of physics as everything else. One of these laws prevents information from being transmitted faster than light. This restriction places a fundamental limit on the speed of any computation that involves spatially separated parts of a processor. Fortunately it may be circumvented by making the processor smaller, but now we hit another limit. When the size of the components approaches atomic dimensions, quantum physics becomes an unavoidable factor. This means that the switching activity is subject to Heisenberg's uncertainty principle, which at first sight looks as if it places an absolute limit on the power of information processing. After all, it's not much good completing a computation in double-quick time if the output is unreliable.

It took the genius of theoretical physicist Richard Feynman to spot that a sin could be turned into a virtue.

To be sure, the intrusion of quantum effects presents severe technological challenges, but it turns out that quantum mechanics also offers a fantastic computational opportunity. The reason for this lies in something called, rather mysteriously, the 'entangled state'. Einstein spotted decades ago that if quantum mechanics is a valid account of microscopic systems, then it implies something truly weird about the nature of reality. To see why, suppose two quantum particles—for example, photons of light—are produced in a single atomic process, and fly apart. The rules of quantum physics require that although the particles may be far away from each other, nevertheless each photon carries a ghostly imprint of the other. By inspecting one photon, we can deduce some properties of the other, distant, photon, without ever directly observing it. The way this information is enfolded in the quantum state is qualitatively quite unlike anything that we might encounter in daily life. It completely defies common sense. In particular, any notion that the two photons possess well-defined properties independently and in advance of interrogation is known to be inconsistent with quantum principles.

The existence of entangled states is not just a theoretical concept: it has been well verified by experiment. The nature of entangled information is subtly different from the normal 'bits' of information encoded in today's digital computers, where all the components are in definite and distinct states—like 'on' or 'off'—at any given time. The opportunity for computation lies not so much with this qualitative difference as with a quantitative one. Put simply, an entangled quantum state can contain more information than could possibly exist in any classical (i.e. non-quantum) state involving the same number of particles. And the quantum advantage escalates with the numbers of particles involved. For a system with even a thousand components, the difference is huge. This invites the idea that one might actually use quantum processes themselves as a means of computation; in other words, carry out the basic computational steps at the atomic or subatomic level.

Though it is easy to spot the potential of quantum computation, implementing it is quite another matter. Entangled states are notoriously fragile. Though simple entangled states have been created in the laboratory, they require great care and unusual conditions. The difficulty is that all quantum systems are continually assailed by disturbances from their surroundings. Experience shows that in almost all realistic cases, quantum entanglement gets hopelessly scrambled by such environmental noise in an exceedingly short time. Some researchers believe this 'decoherence' severely limits the scope of quantum computation. Certainly, if quantum information processing is ever to progress beyond the trivial, a way to protect the physical system from external disturbance (or to compensate in some way for it) must be an urgent priority.

Suppose that these formidable technical obstacles can be overcome, then what use is a quantum computer anyway? Here's one use. Computer users are well aware that some problems are simply too tough for a conventional computer, however powerful it may be. These computationally intractable problems arise because the time required to work through them exponentiates (or worse) with the scale of the problem. A famous case concerns the factorisation of large numbers into primes. Even a supercomputer would take longer than the age of the universe to finish the task for a sixty-four digit number. In principle, however, a quantum computer could crack the same problem in pretty short order. The secret of the quantum computer's incredible power lies in the subtle nature of quantum states. One graphic (though still contentious) way of thinking about quantum uncertainty is to suppose that all possible outcomes encompassed within the range of uncertainty are actually realised in parallel realities. This is the famous 'many universes' interpretation of quantum mechanics. If, for example, a quantum switch has a 50–50 chance of being on or off, then according to the many universes view, there will be two worlds, one with the switch on, another with it off. The two worlds co-exist in a sort of hybrid reality, like two movies projected onto the same screen. In more general situations, this duplication extends to vast—even infinite—numbers of alternatives. Viewed this way, quantum physics provides naturally for massive parallel computation, only instead of the parallel processing being conducted in adjacent parts of the computer, as is the case with existing machines, they are carried out in adjacent universes!

It would be wrong to conclude that the quest for a quantum computer is driven solely by the desire to factor large numbers and perform similar numerical feats. The subject is studied as much for the physical and philosophical insights it provides. Since Newton, scientists have tended to think of the universe as a gigantic clockwork mechanism. Increasingly, however, this mechanistic and materialistic picture of the physical world is being replaced by a new notion: nature as a computational process. Many scientists now see the world as a vast informational frolic, in which the hardware—the actual bits and pieces of the universe, like subatomic particles—are secondary to the entagled information they carry and the complex interactions they engage in. The Oxford theoretical physicist David Deutsch has stressed that the logical and rational structure of the universe cannot be separated from the laws it obeys. What can or cannot be computed will be determined by the form of those physical laws. Thus, computation in a quantum universe is completely different from computation in a classical universe. Deutsch believes that a quantum universe has the deep property that it may be perfectly simulated by only part of itself. That is, a quantum computer has the power (in principle!) to create a virtual reality simulation indistinguishable from the original. If this is so, then the universe has a type of self-reflective quality suggestive of a fundamental principle of existence.

These, and other heady topics, form the subject of this breathtaking book by Gerard Milburn. As one of the world's leading researchers in the foundations of quantum mechanics, Milburn has already achieved international acclaim with his earlier book *Quantum Technology*. There are very few professional quantum physicists able to communicate clearly and without exaggeration the astonishing consequences of quantum weirdness. His careful and thorough account of how entanglement leads to observations that are totally impossible in a classical world is the best I have read. Reporting back from the cutting edge of fundamental research, Milburn gives us an up-to-date account of the subject of quantum manipulation, a field where experimenters probe individual atoms, where entangled states are used to clone subatomic particles, and where cryptography takes on a whole new meaning. Emerging from this sizzling account is a sense of momentous advances tantalisingly glimpsed. Quantum computation is the nexus at which 'software' concepts like information, randomness and order tangle with 'hardware' concepts such as matter and force. I am convinced that in the search for a unified theory of nature, in which the deepest level of reality would be exposed, the study of quantum computation will mark the beginning of a new era.

Paul Davies
Adelaide

CONTENTS

PREFACE

As the century and the millennium approaches an end, I have noted, with growing unease, the number of books appearing in my local book store with titles beginning, *The End of* . . . I passed by *The End of History* with mild amusement, but when *The End of Science* recently made an appearance, I was not amused. Perhaps I should have been, as at first sight the title looks ridiculous. Science end? Surely not just yet. Look around you. Does it look like we have solved every problem to which Science might legitimately be applied? There is plenty of Science to be done just cleaning up the mess left by earlier misplaced scientific and technological endeavours. Global warming, for example.

What the author had in mind, however, was a kind of 'grand science'. In the case of physics, so goes the claim, we now have the general plan of physical reality. Sure, there are details to be worked out, but there will be no big surprises like relativity or quantum theory. It is possible that this is true, however, those of us actually doing physics would almost all agree that there is little evidence for it. One only has to consider that our two grandest theories— quantum theory and general relativity—simply do not fit together. There are some promising mathematical proposals to get them out of one more general theory, but so far none of these proposals come close to solving any of the big problems we are facing in trying to combine gravity and the quantum principle. It is true that these new theories do go a long way to getting gravity out of the quantum principle, but really they only solve a number of very limited questions. Of course any science will eventually answer a finite set of questions. That simply requires time, money and technical expertise. The real issue is asking questions in the first place. It is the question that signifies the creative act in science. Science will only come to an end when we cease putting questions to nature in the form of experiment. A big surprise only requires the right question.

This book is about the quantum principle, a very surprising discovery about the nature of physical reality, painfully extracted by a number of physicists in the early part of this century. The principle is this: physical reality is irreducibly random. At its heart the cosmos is a massive collection of dice throws, an endless coin-toss. But how can anything so beautifully ordered and lawful as the universe arise from such an apparently lawless principle? That is the essential mystery lurking behind the quantum principle. The answer lies in the deeply puzzling way nature tosses the coins.

Everything we know about the physical world can be put in terms of answers to a sequence of 'yes/no' questions. At the quantum level, those yes/no results usually fall like the toss of a coin. But the odds are not at all what we might have guessed. At a deeper level the answer to every yes/no question is determined by a qubit, the fundamental quantum description of a yes/no question. In a single qubit we see the face of infinity. For a coin-toss we can only ever ask a single question. But for a qubit we can go on asking questions forever and it will keep supplying yes/no answers. The quantum world is a world of endless surprise. In this I see a warning for all those who would claim that we will soon have all the answers. The qubit is nature's way of hiding infinity in a coin-toss. All of the most puzzling features of the quantum theory stem from this. If reality is to be both irreducibly random, yet apparently intelligible, the qubit is probably the only way to roll the dice. There is no mind behind the dice game of reality. The apparent intelligibility of the universe is constructed, bit by bit, from entangled qubits, the smart dice of reality.

ACKNOWLEDGMENTS

I would like to thank Damien Pope for carefully reading a number of early drafts of this book and making some valuable suggestions. Thanks to Ike Chuang for permission to use his Schroedinger cat Wigner function on the front cover and also to Dave Wineland for permission to use his ion trap drawing in Figure 6.2. Thanks also to Cathy Holmes for drawing Figure 4.3. I would like to express my gratitude to Bill Munro and Howard Wiseman for many discussions on the nature of quantum entanglement and helping to clarify some aspects of Chapter 3. None the less, quantum entanglement remains a puzzle to us all.

CHAPTER ONE

THE QUANTUM PRINCIPLE: INEXHAUSTIBLE UNCERTAINTY

What there is and how we know it

What is the stuff of creation? Is all reality simply matter in motion or something else entirely, a complex concert of vibrations in a primordial field? Is the world we see but a gross simplification of an underlying cosmic chaos, or is there a fundamental simplicity and harmony at its heart? Is what we see and experience the unfolding of a single simple rule repeated over and over again at every level of physical reality, or are there endless rules each with its own domain of application? Just what is the world made of, and can we know it? These are ancient questions, inevitable as the fact of our own consciousness. As this century draws to a close we think we know the answers to some. Quantum theory is our most fundamental and successful description of what there is. Yet the view of reality it presents is so bizarre and so at variance with common sense, that after almost a century we are still debating just what it means. The quantum principle appears to apply to all reality, yet we still cannot agree on how the world we see can be derived from a principle at once so simple and so perplexing.

The quantum principle is this: physical reality is irreducibly random, but random in a way we could never have expected. Reality exhibits a randomness so constrained, by an as yet undiscovered principle, that the odds with which it deals confound anyone who studies it. The fact of this

1

randomness is deeply shocking and acknowledged as such by the originators of the theory. Why should the universe, at its most fundamental level, present us with an inexhaustible source of uncertainty?

A perusal of the writings of quantum physicists from the original discoverers to the present day, continually exposes confusion, wonder and disbelief at a world so built. Einstein complains in a letter to Max Born that:

> Quantum mechanics is very impressive. But an inner voice tells me that it is not yet the real thing. The theory produces a great deal but hardly brings us closer to the secrets of the old one.

Richard Feynman, who received the Nobel Prize in 1965 for the quantum theory of light, put it like this:

> . . . we always have had (secret, secret, close the doors!) . . . a great deal of difficulty in understanding the world view that quantum mechanics represents. At least I do, because I'm an old enough man that I haven't got to the point that this stuff is obvious to me. Okay, I still get nervous with it.[1]

Yet the quantum theory is the most successful physical theory ever. It makes predictions which have been tested to an unprecedented level of accuracy. Its most counterintuitive predictions continue to be verified in ever more sophisticated tests. It gives us an understanding of the physical world upon which most of modern technology is based, from the marvels of microelectronics to supermarket laser scanners. It will underpin, to an even greater degree, the high technology of the new millennium, a quantum technology.[2] There simply cannot be any doubt that nature is built according to the quantum principle. Yet despite this undisputed success, debate continues. How can the apparent order of the world be built on irreducible randomness? The answer is not yet known, but the key lies in the peculiar nature of quantum randomness.

A quantum coin-toss

The year 1905 was a big year for Einstein. In that year he published the special theory of relativity, and gave an explanation for the peculiar phenomenon of Brownian motion, vanquishing at last the few remaining doubters of the atomic hypothesis. In that year he also explained the photoelectric effect, ushering in the modern understanding of light and the idea of a photon. The nature of light has been a perennial problem for physicists. Newton supposed it to be composed of a stream of small particles. His model did a good job of explaining reflection and refraction, but failed when it came to effects such as diffraction, the slight bending of light rays around sharp edges. In the nineteenth century, a new wave model for light achieved dominance, easily explaining both reflection and refraction as well as diffraction and interference. James Clerk Maxwell achieved the final synthesis in the wave theory of light when he proved that light was a form of electromagnetic radiation. But, even as Maxwell's great synthesis was achieving orthodoxy, a new phenomenon was raising doubts on the veracity of the electromagnetic wave theory of light. This phenomenon, called the photoelectric effect, was the property of certain metals to develop an electric charge when exposed to ultraviolet light. The photoelectric effect now forms the basis of countless everyday devices, from supermarket laser scanners to pollution detectors in large industrial chimneys.

Every attempt to explain the details of this phenomenon in terms of Maxwell's wave theory failed totally. Building on earlier work by Max Planck to describe the light emitted by hot objects, Einstein, in 1905, hit upon a new model of light that easily explained all features of the photoelectric effect. The idea was simple and at first sight a return to the Newtonian model of light. To explain the photoelectric effect, said Einstein, think of light as a rain of infinitesimal identical 'packets' of energy, called photons.

In some respects a photon is like a particle. They carry both momentum and energy and can collide with real

3

particles, such as electrons, much like little billiard balls. This is how the photoelectric effect is explained. A photon of the right energy can knock an electron free from the surface of a metal, leaving behind a net positive charge. But there are some big differences. First, photons are created and destroyed whenever light is emitted or absorbed, without any significant change in the mass of either the emitter or the absorber. Second, photons in a vacuum do not interact at all, but simply pass right through each other. Finally, the energy carried by each little packet is fixed once and for all by the colour of the light, blue light photons having a greater energy than red light photons.

The modern theory of light explains how it can be described both as an infinitesimal particle-like object, yet at the same time exhibit all the features of a wave. The explanation takes us straight to the heart of the quantum principle, for the photon is the quintessential quantum object. The explanation is quite simple though more than a little surprising. Let me take you through it. Consider a very simple optical device: the half-silvered mirror. This device, which I will call a beam-splitter, simply splits a light beam into two components of equal intensity, without diminishing the overall intensity and without changing the colour of the light. If we take an input beam and direct it towards a beam-splitter, and measure the intensity in each output direction, we find it is split equally.

To measure the intensity of light we can use the photoelectric effect. The more electrons ejected, the more photons per second are incident from the light. The ejected electrons can then be drawn off to form an electric current, so that the greater the intensity of the light, the greater the current. If the intensity of the light is very low, we can set up the photoelectric devices to count individual photons. In a practical device we will miss quite a few, but we will assume that in principle we can make perfect photon counters using the photoelectric effect.

A wave theory of light has little difficulty explaining what happens at a beam-splitter. The wave is simply split

Figure 1.1 The behaviour of light at a beam-splitter. Each photon counter will register an equal number of photons on average, at each of the detectors.

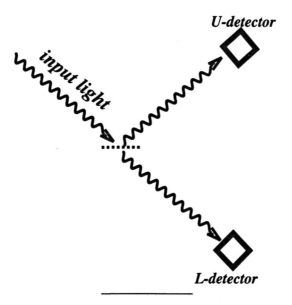

so that a wave of half the intensity is transmitted and a wave of half the intensity is reflected. The two detectors would measure equal intensities. This explanation is fine so long as the light is sufficiently intense. However, as the intensity is decreased, we begin to see the granular nature of light.

If we put a photon counter at each of the two output directions from a beam-splitter and turn the input intensity right down, what do we see? The measured intensity of the light begins to fluctuate. Sometimes the L-detector counts a few more photons than the U-detector and vice versa, although these fluctuations may be very small compared to the total photon count at each detector. As we continue to turn the intensity down, the fluctuations become very significant. At the single photon level, when in any given time

5

interval we only count a single photon, we reach a situation in which we either count no photon at all, or we count a single photon at the U-detector and none at the L-detector, or we count a single photon at the L-detector and none at the U-detector. Which detector actually registers the photon is as random as the toss of a coin. Even though we know everything there is to know about the incoming light, we cannot predict which detector will register the photon. If we repeat the experiment over and over again at the single photon level, we find each detector counts roughly half of the total number of photons counted. We have reached the irreducible randomness of the quantum.

To explain this experiment we can propose that when a single photon encounters a beam-splitter it will be reflected or transmitted with equal probability (I will assume the beam-splitter does not absorb any photons). No matter how much we know about the incoming light and the beam-splitter we cannot and need not say more than this. All we can give is the odds for reflection versus transmission. In fact we do have an exact theory of light and beam-splitters. It is the quantum theory, and for this experiment it only gives the odds for reflection and transmission. Nothing more exact than that. You may be thinking that surely we should try to learn just a bit more about photons and beam-splitters so as to predict what any individual photon will actually do. Perhaps photons come in two kinds. One of these is the kind which gets reflected at beam-splitters and the other is the kind that gets transmitted at beam-splitters. Perhaps each photon carries some kind of instruction as to what to do when it encounters a beam-splitter, some kind of 'gene' which tells it how to behave in the circumstances. Such a hidden variable would, of course, be a perfectly valid way of explaining the apparent randomness of photons at beam-splitters. If the quantum theory does not enable us to say what this hidden variable is, that is too bad for quantum theory. Perhaps someone should just try a bit harder.[3]

This is a perfectly valid response to any kind of

randomness. It is based on the idea that nothing is really random. That somewhere there is a hidden archive, which if accessed, could be used to predict with certainty what would happen in every case. This is certainly the classical ideal of physical explanation. The book of nature is written once and for all. Every element of physical reality has its marching orders. In this classical picture, each experiment is an attempt to set up conditions to reveal the particular rule which is supposed to apply in that case. Every particle in the experiment does what it does by virtue of its hidden instruction sets, its intrinsic nature. Any observed randomness simply reflects our lack of knowledge of the hidden archive.

If hidden instruction sets exist, what properties must they have? If all photons come with a gene for reflection and a different gene for transmission at a beam-splitter, what kind of world would we see? Postulating hidden instruction sets to explain randomness has important consequences. The hidden instructions had better all be logically consistent with one another. If this were not the case, experimental results would lead to contradictions— such a world is an impossibility. A photon carrying the reflection-gene would always be reflected from identical beam-splitters. The instruction sets should consistently account for every experiment, not only those experiments which are explained by a particular hidden variable, but every conceivable experiment. A photon carrying the reflection-gene for beam-splitters, when subjected to some other optical experiment, cannot lead to results that can only be explained by a hidden variable that is inconsistent with a reflection-gene. The hidden instruction sets must account for all experiments that might conceivably be done in the future. If this were not the case, the past would be no guide to the future, or the future might not be consistent with the past. We may be able to accept that instructions sets can change in time, so long as they do so in an entirely predictable and deterministic way.

A more subtle feature of hidden instruction sets, and

7

possibly one which we can dispense with, is required by the special theory of relativity. We assume that the results of a measurement on some aspect of physical reality is explained by the instruction sets of only those elements of reality involved in the experiment. For example, whether or not a given photon is reflected at a beam-splitter should not depend on what happens to another photon in a distant laboratory. Likewise, we would not expect that whether or not a given photon will be reflected will depend on what happened to a photon at that same beam-splitter, any time in the future. These are referred to as *locality* assumptions. We do not exclude the possibility that measurements in one place and time are influenced by measurements in another place or time, so long as some possible signal can connect the hidden variables involved in each experiment. This signal cannot travel faster than the speed of light. Local hidden variable theories have been extensively investigated as a source of randomness in quantum mechanics, but have thus far failed to account for experimental results. We shall return to this subject in the next chapter.

There is a simple experiment which suggests that a simple hidden variable, or reflection-gene, to determine whether or not a given photon is reflected is not possible. Suppose we set up a sequence of identical beam-splitters as in Figure 1.2. A photon incident on the first beam-splitter will be either reflected or transmitted. Suppose this is determined by an unknown hidden variable which can take one of two values. One value will determine that this photon is reflected, and the other value will determine that this photon is transmitted. If it is the kind of photon that is reflected, it will be reflected at the first beam-splitter, and that is an end of the line for that photon. We get a single count at the first detector and no other detector will fire. However, if it is the kind of photon that is transmitted, it will be transmitted by the first beam-splitter and every subsequent beam-splitter. In that case no detector will ever fire. Quantum mechanics predicts that while both these outcomes are possible they are not the only possibility. A photon can be

Figure 1.2 A single photon encounters a sequence of beam-splitters. The quantum theory predicts that every encounter with a beam-splitter is totally random. Thus, any of the detectors may record the photon. If the propensity to be reflected or transmitted was a property of the photon and determined by a hidden variable R or T, an R-type photon would be detected with certainty at the very first detector in every run.

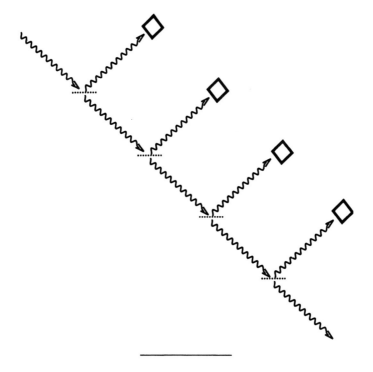

transmitted by any number of beam-splitters and still have a non-zero chance of eventually being reflected. In the quantum theory, there is a chance for any of the detectors to fire. This is easily explained if we assume that at every encounter with a beam-splitter, a photon makes a random choice to be reflected or transmitted, regardless of its previous history.

Every encounter is another coin-toss. The fate of the photon at every beam-splitter is always uncertain. The uncertainty is inexhaustible.

We could simply accept the fact that when photons encounter beam-splitters they behave in an entirely unpredictable way. At each encounter, there is an equal chance a photon will be reflected or transmitted. This will still enable the photon picture to be consistent with a wave picture. If the intensity of the light is very large, roughly half the photons will be reflected and half transmitted. For all practical purposes this would result in two beams of equal intensity, just as the wave theory predicts. This is indeed what the quantum theory of beam-splitters shows. But there is more to it than that, as will become clear when we consider multiple encounters with beam-splitters.

Suppose we take a photon through two beam-splitters in succession, as in Figure 1.3. An incoming light beam is split equally and each component of the beam travels along two separate paths, called the upper path (U-path) and the lower path (L-path) until again falling on a beam-splitter. After the beam-splitters, the two output beams each fall on a photon counting device, labelled L and U in Figure 1.3.

We can adjust the path lengths inside the optical device by moving the top mirror a little, as indicated by the double-headed arrow in Figure 1.3. We can adjust the setting of this mirror so that only detector L registers any intensity, that is, there is no output beam directed towards the U-detector. Adjusting the paths further, we reach a situation in which equal intensity is measured in the U-detector and the L-detector (as shown in Figure 1.3). Adjusting the paths further still we reach a situation in which no intensity is measured at the L-detector; all the light emerges towards the U-detector.

The wave theory of light easily explains these observations in terms of interference. Suppose we know everything there is to know about the source of the light, its intensity, its colour and its polarisation. The incoming light wave is split equally into two waves of half the

Figure 1.3 A simple optical device to illustrate the wave nature of light. Incoming light is split equally at a half-silvered mirror, and travels along two separate paths, before being recombined at another half-silvered mirror. Finally the two output beams fall upon photon counters. The number of photons counted per second is proportional to the intensity of the light. When operated at very low intensity so that on average only one photon at a time is counted, the operation of the device is explained by the quantum theory.

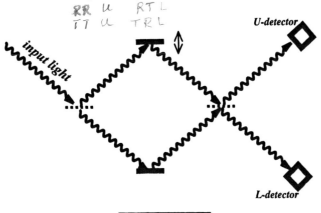

amplitude. The two waves recombine at the final beam-splitter, but now the two output waves can interfere with each other. There are two possible output beams. A beam-splitter, however, combines the waves differently for each of these directions. For one direction it simply adds the waves together. For the other direction it first inverts one wave, that is, it takes a wave crest and turns it into a trough. If a crest from one wave exactly matches the crest from the other wave, there is a perfect recombination of the two waves. When this happens all the light exits the device along one beam. For the beam in the other direction, however, a crest is combined with a trough and there is complete cancellation. No light comes out in this direction. By adjusting the path length in each arm of the device we can arrange for only partial addition or cancellation of a

11

crest and a trough. In this way we can ensure that roughly half the light intensity is measured in each output beam.

The explanation in terms of waves is quite straightforward and is easily verified in the laboratory. Interference is the defining characteristic of a wave. How could we possibly account for this phenomenon if light was a particle? Usually we don't need to. Typically we do experiments with so much light that we never detect individual photons. The detectors simply register the average intensity, and do not record individual photon counts. But we can do the experiment with individual photons.

Suppose we now begin to turn down the intensity of the source of the input light. Keep turning it down so that on average the photon detectors at the output measure only a single count in any time interval. Now what do we see? Suppose we started with the situation in which all the light was detected at only one output beam, say the L-detector. As we turn the intensity of the input light down we see less and less intensity at this detector, but we never see any photons counted at the U-detector. At the very lowest intensity, we see single photons arriving at the L-detector, randomly in time, but we never see a single count at the other detector. That result is at least consistent with the wave theory.

Now set up the paths so that we begin with a situation in which half the light emerges in each output beam. Again, we turn the intensity of the input light down until only a single count is recorded at either detector in any time interval. At the single photon level, when in any given time interval we only count a single photon, we reach a situation in which we either count no photon at all, or we count a single photon at the U-detector and none at the L-detector, or we count a single photon at the L-detector and none at the U-detector. Which detector actually registers the photon is as random as the toss of a coin. Even though we know everything there is to know about the incoming light and all the optical devices in the system, we cannot predict which detector will register the photon. Again, we reach a case of irreducible uncertainty.

Suppose each photon does carry a gene, or hidden variable which determines if it will be reflected or transmitted at a beam-splitter. In Figure 1.3 we can see that if it is reflected at the first beam-splitter, it will be reflected at the second beam-splitter, resulting in certain detection at the U-detector. If it is transmitted at the first beam-splitter it will be transmitted at the second beam-splitter, again resulting in certain detection at the U-detector. Thus if photons carry genes for beam-splitters, the light will always be detected at the U-detector. This would be the case regardless of the path length difference between the two arms of the device. Clearly this prediction is not correct.

Suppose instead we accept the fact that the behaviour of a photon at a beam-splitter is entirely random, and that there is no hidden instruction set to explain this result further. Each encounter with a beam-splitter is a coin toss. Now comes the crunch—this explanation is inadequate as well. There are four possible events to describe what happens to a photon between input and output: photons can be reflected at both beam-splitters; or they can be transmitted at both beam-splitters; or they can be reflected at the first beam-splitter and transmitted at the second; or finally, they can be transmitted at the first beam-splitter and reflected at the second. For ease of reference let me label these four cases as RR, TT, RT and TR. Note that RR and TT both result in certain detection at U, while RT and TR both result in certain detection at L.

The experiment is like tossing a coin twice in succession, with one side labelled R and the other side labelled T. What we need to determine is the chance of a photon being counted at U or L. To do this we need a couple of rules for combining probabilities. The first rule was invented by Laplace in the eighteenth century and goes by the rather ironic sounding name of 'Laplace's rule of insufficient reason'. It says that, if we don't know any better, a set of alternative outcomes should be assumed to have equal probability. In this way the toss of a single coin can be assumed to have equal probabilities for each of the two possible results. The probability for a head is 50 per cent, as is the probability of a tail. If we toss the coin twice we have four possible outcomes, so that each outcome will have a probability of 25 per cent. In our experiment this means each of the histories RR, TT, RT and TR, will each have a probability of 25 per cent. The second rule cannot be traced to any particular individual, but was in common use in the nineteenth century, particularly in the work of Bayes. I will thus call it Bayes' rule. This rule tells us how to combine probabilities. It says that if an event can happen in two or more ways, add the probability for each way, considered separately. In our experiment a detection at U can happen in two ways, RR or TT, so the probability for detection at U is 50 per cent. Thus we predict that if a photon responds to a beam-splitter like a coin-toss, in a long run of experiments on single photons, we will detect equal numbers at U and L, independent of the path length difference in the device. Again this fails to account for the experiment. So now our difficulty is apparent. At low intensities we only ever count one photon at a time. Each photon behaves like a coin-toss at each beam-splitter, yet the outcome of the experiment must depend on the path length difference, if it is to be consistent with observations at large intensities, and thus consistent with the wave theory of light. How are we to marry irreducible randomness with interference?

The key is to find a way for the probabilities to depend on the path length difference. We still must keep the idea that at each beam-splitter the photon has an equal chance to be reflected or transmitted. But when two successive encounters with a beam-splitter are required, the probability for the combined events must depend on path length difference. The quantum theory manages the explanation by replacing both Laplace's rule and Bayes' rule. First, probabilities are not fundamental, but are determined at a deeper level by a probability amplitude. This is not like a hidden variable, however, as knowing the probability amplitude only enables you to calculate the probability for an event. It still does not enable you to predict with certainty which particular event is realised. Second, we replace Bayes' rule for combining probabilities by a new rule for combining probability amplitudes.

Given a probability amplitude, how do we calculate a probability? A probability amplitude depends on two real numbers. Given these numbers we get the probability by squaring each of them and adding. Now that may sound familiar. What else depends on the sum of the squares of two numbers? Yes, Pythagoras' theorem, which tells us how to calculate the length of the hypotenuse of a right angle triangle from the length of the other two sides. What a surprise! The world is irreducibly random and the odds turn out to be determined by a formula discovered by an ancient Greek sage over two millennia ago. Perhaps quantum randomness is geometry at heart?

To picture a probability amplitude we can draw a little arrow as the hypotenuse of a right angle triangle. The length squared of this arrow is the probability for the corresponding event. However, what the quantum theory gives us is the particular right angle triangle for that hypotenuse. I have indicated this pictorially in Figure 1.4.

Of course there are infinitely many right angle triangles we can draw for a given hypotenuse (the absolute direction is unimportant). How do we know which one corresponds to a given event? The answer is, it depends on the physical situation, and the quantum theory tells us how to assign the right probability amplitudes for each physical situation. As an example, let us look at the case of a photon at a beam-splitter. To determine what happens at a beam-splitter we need two probability amplitudes, one for reflection and one for transmission. We thus need two right angle triangles. However each will have an hypotenuse of the same length, as reflection and transmission are equally likely. As the length *squared* is the probability, we need to have two triangles each with an hypotenuse of length $\frac{1}{\sqrt{2}}$. For a single

Figure 1.4 A pictorial representation in terms of arrows of two different probability amplitudes that correspond to the same probability. One probability amplitude is $(\frac{1}{2}, \frac{1}{2})$ while the other is $(-\frac{1}{2}, -\frac{1}{2})$. Each amplitude is represented as an arrow with the tail at the origin, and the head at the coordinates corresponding to the relevant amplitude.

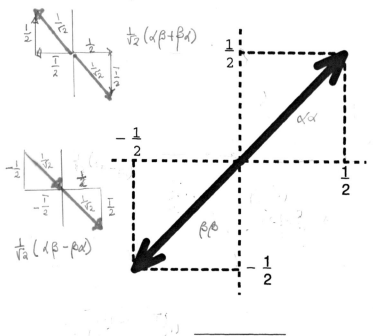

beam-splitter, it doesn't really matter how we draw the triangles, just so long as they each have an hypotenuse of this length.

If only the length squared of the hypotenuse matters, why bother with the probability amplitude at all? The answer comes when we try to combine events. In classical probability theory, Bayes' rule tells us how to combine the odds for an event that can happen in various ways. In quantum theory, this is replaced with a rule for combining probability amplitudes. I will call this Feynman's rule after

16

Richard P. Feynman, a famous American physicist who gave us this simple picture of the quantum theory.[4] Feynman's rule says, if an event can happen in two or more *indistinguishable* ways the probability amplitude for that event is the 'sum' of the probability amplitudes for each way, considered separately. The final probability of the event is then obtained by the sum of the squares of the two numbers describing the resultant probability amplitude. But if a probability amplitude is determined by two numbers, how do we add probability amplitudes?

The way to add probability amplitudes is to first group the two numbers involved in some order. For example, the probability amplitude for reflection at a beam-splitter might be written as $(\frac{1}{\sqrt{2}}, 0)$, or it might be written as $(\frac{1}{2}, -\frac{1}{2})$. In both cases, the sum of the squares of the components are $\frac{1}{2}$ as required, but the components need not be positive. Once a particular choice is fixed for one probability amplitude, all the other probability amplitudes are fixed. There is a simple geometric way to keep track of this. We first draw a set of perpendicular axes as in Figure 1.4. The probability amplitude is then represented by an arrow. The projection of this arrow on each of the two axes determines the two numbers which make up the probability amplitude. If we only have a single probability amplitude, it does not matter in which direction we draw the arrow. However, if we need to combine probability amplitudes, we must be very careful about the *angle between* arrows for the corresponding probability amplitudes. In quantum physics we can adjust these relative angles to change the odds for a given event.

Once we have fixed a set of arrows corresponding to a set of probability amplitudes, we add the probability amplitudes by adding the corresponding coordinates of each amplitude. In Figure 1.4, for example, we have two probability amplitudes $(\frac{1}{2}, \frac{1}{2})$ and $(-\frac{1}{2}, -\frac{1}{2})$. The sum of these is $(0, 0)$. The sum of the squares of this amplitude is, of course, zero. The event thus described will never occur. This is despite the fact that the probability for each of the

17

two constituent events is not zero, but in fact given by $\frac{1}{2}$. Because we are adding probability amplitudes like arrows, the probability for the combined event can be zero, even though the probability of the constituent events is not zero.

As an example let us look again at the two beam-splitter devices in Figure 1.3. Suppose a photon is detected at the U-detector. There are two ways this can be achieved, corresponding to what happens to a photon at each beam-splitter: both photons can be reflected (RR) or both photons can be transmitted (TT). Likewise there are two ways that a photon can be detected at the L-detector, RT or TR. The probability for a U-detection or an L-detection can each happen in two indistinguishable ways. Consider the U-detection case. To determine the probability for this we need to know the probability amplitudes for RR and TT and then add them in accordance with Feynman's rule. The results are summarised in Figure 1.5.

We see that in Figure 1.5 (a) the probability amplitudes for RT and TR add together perfectly, while the amplitudes for RR and TT cancel exactly. This would mean the probability to detect a photon at L is one, while the probability to detect a photon at U is zero. In Figure 1.5 (b) we see an intermediate situation in which both detection events occur with equal probability.

The quantum theory of our beam-splitter device is complete when we postulate that by changing the path length in each arm, we change the angle between the probability amplitudes for each history. If we set up the device so as to give certain detection in L and slowly change the path length, the angle between all probability amplitudes changes. The change of angle for RR and TT are equal and opposite, as are those for RT and TR. Eventually we reach the situation in Figure 1.5 (b) where a photon is equally likely to be detected at U as L. If we keep going we get to a situation where the amplitudes for detection at L cancel exactly, while those for detection at U add exactly, in which case a photon is detected at U with certainty.

By allowing the angle between probability amplitudes

Figure 1.5 A pictorial representation of the probability amplitudes for the various photon histories that lead to detection at one or the other of the detectors in Figure 1.3. In (a) the path length in the device is adjusted so that the amplitudes for the histories RR and TT cancel exactly, making detection at L certain. In (b) the path lengths are adjusted to make detection at each detector equally likely.

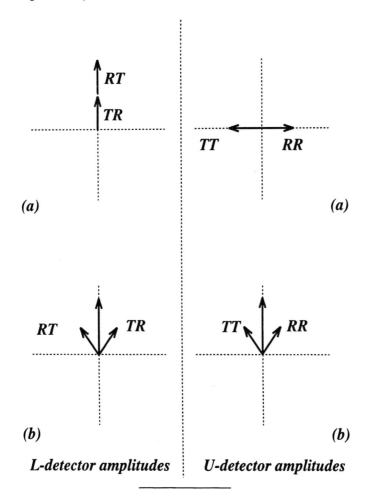

L-detector amplitudes   *U-detector amplitudes*

19

for the history of a single photon to depend on path length, we can easily explain both what happens at the level of a single photon and what happens for many photons passing though the device, that is, for more intense light. Each photon behaves independently of every other and is detected at an output with a probability determined by the relevant probability amplitude. As we vary the path length difference, the final detection rates will be exactly as expected for a wave theory of light.

Photons do behave in an irreducibly random way at a beam-splitter. But the rules for calculating the odds for multiple encounters with beam-splitters are radically different from ordinary probability. The world described by quantum theory is irreducibly random, but the odds are constrained to follow the rules of probability amplitudes. Why should it be the case that an irreducibly random event be constrained by just these rules? Why cannot multiple events be described by ordinary probability? No-one knows the answer to that question. However, I suspect the answer is connected with the answer to another question. In a world which is at its heart irreducibly random, how is order possible at all? But this is pure speculation. The real answer, as always, lies just beyond a new experimental result. Perhaps we should follow Feynman's advice:

> Do not keep saying to yourself, if you can possibly avoid it, 'but how can it be like that' because you will get 'down the drain', into a blind alley from which nobody has yet escaped. Nobody knows how it can be like that.[5]

Polarisation: a new handle for photons

A filter is a simple optical device that is used by many people as a fashion accessory. Sunglasses filter rays of sunlight, making outdoor activities in Australia rather more comfortable. Light can be filtered by selectively absorbing light with different properties. Some filters act on colour. They pass light rays of some colour ranges and absorb the rest. Other filters simply act to reduce the intensity of light.

A very important property of light, however, is used in a polarisation filter.

When the great synthesis of electricity and magnetism was achieved in the work of James Clerk Maxwell in the late nineteenth century, light was shown to be a self-sustained wave of electric and magnetic fields. Waves of any kind transport energy from a source to a receiver, without transferring matter. In a vacuum, the energy carried by a light wave is transferred at the speed of light and the oscillating electric and magnetic forces that make up the wave are always directed in a plane perpendicular to the direction of energy transport. Furthermore the electric force is always perpendicular to the magnetic force, so we can simply talk about the electric force alone. The line along which the electric force is directed may be fixed, or it may continuously change in time. The direction of this line is called the polarisation of the light. If the line does not change its direction as the wave moves, the light is plane polarised. If the line slowly rotates through 360 degrees as the wave propagates, the light is circularly polarised. The idea of polarisation is depicted in Figure 1.6.

Sunlight is a tangle of light waves with many different colours and polarisations. A polarisation filter is a partially transparent material which only transmits light of a particular polarisation. If all polarisations are present in sunlight, a polarisation filter absorbs most of the light and transmits only light waves which have a polarisation parallel to some particular orientation of the filter. We can rotate the filter and select out different polarised light waves.

Consider the simple experiment shown in Figure 1.7. A beam of sunlight is first passed through a polarisation filter to select out one particular polarisation. It is then passed through a second polarisation filter. As we rotate the second filter, less and less light gets through, as the second filter is selecting out a different polarisation to that selected by the first filter. At some angle no light gets through at all. In this case we say the polarisers are crossed. Suppose we say the first filter transmits vertically polarised

21

Figure 1.6 In the nineteenth century Maxwell showed that light is a self-sustained wave of electric and magnetic forces. If the light is polarised, the electric force, though varying in time, always points along a single line. In the diagram this is the line in the vertical (V) direction and the corresponding light beam is vertically polarised.

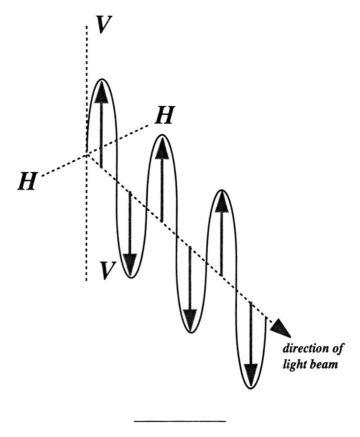

direction of light beam

light. The crossed filter will only transmit horizontally polarised light. The light is vertically polarised after the first filter. It cannot get through the second filter as that filter only passes horizontal polarisation. The two possibilities, horizontal and vertical polarisation, are mutually exclusive and thus good candidate properties for a coin-toss.

Some materials can change the polarisation of light. A solution of sugar is an example. By itself, a sugar solution is completely transparent. It absorbs no light at all. Suppose we place a beaker of sugar solution between the crossed polarisers. Suddenly some light is passed by the second filter. The sugar solution has changed the polarisation of the light transmitted by the first filter. In fact, the sugar rotates the polarisation of the light transmitted by the first polariser. This is equivalent to removing the solution and rotating the second polariser. A number of devices can be used that rotate the polarisation by a fixed angle. It is even possible to obtain devices that rotate the polarisation by any given amount at the turn of a knob.

Some materials can even act as polarisation-sensitive beam-splitters. One polarisation will be transmitted and the mutually exclusive alternative will be reflected. We may, for example, set up the material so that vertical polarisation is transmitted and horizontal polarisation is reflected.

Polarisation would appear to be a property so closely tied to the wave nature of light that at first sight it would seem to make no sense at all to talk about the polarisation of a single photon. However, a simple experiment shows that, in fact, we must be able to talk about the polarisation of a single photon. All we need to do is to turn the intensity down so that on average we only count single photons.

Consider the example of Figure 1.7, but replace sunlight with a laser source, the intensity of which can be carefully controlled. Let us put a single polarisation filter between the source and a single photon detector. Any light that reaches the detector must have a definite polarisation, as the filter only transmits light of one polarisation. If the intensity is turned down so that we only count one photon at a time, we can continue to say that each and every photon has the same, definite polarisation. Suppose we say that for a particular orientation of the polarisers, all the photons that are transmitted are vertically polarised photons.

Figure 1.7 Sunlight is first passed through a polarisation filter aligned to reject all but one direction of polarisation. This light then passes through a sugar solution. Finally it passes through a second polarisation filter aligned with the first. If the sugar solution is absent, all the light that gets through the first filter gets through the second filter. If the sugar is sufficiently concentrated it will rotate the probability amplitude for photon polarisation by 90 degrees, in which case the probability for it to be transmitted by the second filter is zero.

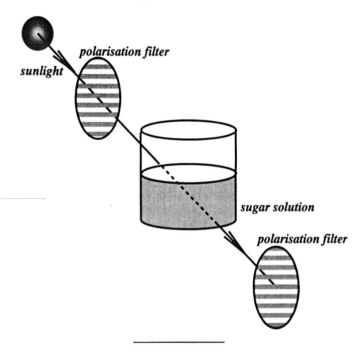

If this line of argument is to be consistent, such photons will not be able to pass a filter crossed to this filter. That is, a vertically polarised photon cannot be transmitted by a filter that only lets through horizontally polarised photons. This must be correct as no light can get through two crossed polarisation filters. A photon is either horizontally polarised or vertically polarised. These are two mutually exclusive alternatives. Of course it doesn't matter if we call the two mutually exclusive possibilities horizontal or vertical. Any two perpendicular directions would do just as well. It looks like there is a coin-toss hiding in the notion of photon polarisation.

It is easy to make a coin-toss out of photon polarisation experiments. Suppose we pass vertically polarised light through a polarisation filter rotated by 45 degrees instead of a fully crossed filter. The experiment shows that roughly half the light gets transmitted. This is very much like a beam-splitter, except there is no reflected beam. Light that is not transmitted is absorbed. Now again turn the intensity of the source down so that we count single photons. What does a single vertically polarised photon do when it encounters a filter rotated at 45 degrees to the vertical? The answer is, it is either absorbed or transmitted with equal probability. This is the only way to ensure that when the intensity is again turned up, half the light will be transmitted. The transmission or absorption of a single polarised photon is a coin-toss. The two mutually exclusive possibilities for photon polarisation lead to a coin-toss, but is this a quantum coin-toss? Are the random results obtained with polarised photons determined at a deeper level by probability amplitudes?

The mutually exclusive possibilities for photon polarisation are, in fact, determined by probability amplitudes, just as reflection and transmission at a beam-splitter. Recall the experiment with the sugar solution. The sugar solution can rotate the polarisation of a light wave. When no sugar is present, no light makes it through the two crossed polarisers. Let us use just enough sugar solution so that roughly half of the intensity of the sunlight gets through the two crossed polarisers, when the sugar is present between them. Suppose we do this experiment with our variable intensity laser source. Now we find that roughly half the photons are transmitted and half absorbed. The sugar solution does not absorb any photons, as the sugar solution is completely transparent, as may be easily verified by removing both

25

polarising filters. However, the sugar solution *can* change the probability amplitudes for the two mutually exclusive directions of polarisation for a single photon. If the probability amplitude for a vertically polarised photon is changed, its subsequent encounter with a crossed polarising filter will be described by different probability for absorption or transmission.

Here is how it works. Suppose we denote a vertically polarised photon by the probability amplitude shown in Figure 1.8. The little arrow is directed vertically in this abstract picture. The probability amplitude for a horizontally polarised photon can be rendered as a horizontal arrow.[6] In numerical terms we write the amplitude of a vertically polarised photon as $(0, 1)$ and a horizontally polarised photon as $(1, 0)$. The sugar solution rotates the amplitude arrows by an angle that depends on the concentration of the solution. If we use just the right concentration, the amplitude for a vertically polarised photon will be rotated by exactly 45 degrees. The photon polarisation amplitude is now neither vertical nor horizontal, but an equal superposition of the two possibilities. The probability amplitude is $(\frac{1}{\sqrt{2}}, \frac{1}{\sqrt{2}})$. If this photon now encounters a polarisation filter at a horizontal orientation, the probability amplitude for transmission is $\frac{1}{\sqrt{2}}$, so that the probability for transmission is $\frac{1}{2}$. The sugar solution can manipulate the probability amplitudes for polarisation, without absorbing the photon.

Seeing in the dark

In the high desert of central New Mexico lies a remarkable town. To reach it one leaves the delightful village of Santa Fe and takes Interstate 25 north a little way before branching west at the Indian pueblo of San Ildefenso. A steep climb to a high mesa, and you find yourself in Los Alamos, a town devoted to the pursuit of a single activity, Science. Los Alamos was brought into existence to create the first atomic bomb. These days, the scientific and engineering activities are more diverse, from projects that seek to

26

Figure 1.8 The probability amplitude for polarisation states may be rotated by physical processes. A sugar solution, for example, can change the polarisation probability amplitude by 45 degrees if the concentration is right. In the figure a photon certain to be detected in the V polarisation state is changed so that it will now be detected in H or V polarisation states with equal probability.

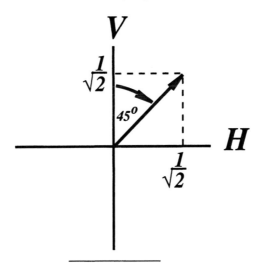

understand how galaxies collide to attempts to bring under control the massive information overload of the human genome project. Paul Kwiat is J.R. Oppenheimer fellow at Los Alamos National Laboratory (LANL). He is pursuing an interesting goal: to use quantum uncertainty to detect objects without touching them, a quantum trick to see in the dark.

Return to the simple optical device of Figure 1.3, again operated one photon at a time. The photon makes a random choice to be reflected or transmitted at the first beam-splitter. The odds for each option we know are determined by a probability amplitude. What would happen if we put some kind of detector just after the first beam-splitter to see which choice was made by the photon? If the photon is reflected it takes the upper path, if it is transmitted it takes the bottom path. These are mutually exclusive possibilities. If we detect a photon on the upper path it must have been reflected at the first beam-splitter. We don't really need a detector on the lower path as well, for if we fail to detect a photon on the upper path it must have been transmitted at the first beam-splitter and will have taken the lower path. Kwiat and his former colleagues at the University of Innsbruck, Harald Weinfurter and Anton Zeilinger, have made this detection event rather dramatic following an idea first proposed by the two Israeli physicists Elitzur and Vaidman. Suppose that we place a light-sensitive bomb in the upper path, perhaps a bomb so sensitive that a single photon will set it off. A photon reflected at the first beam-splitter will certainly result in the destruction of the laboratory.

Kwiat and his colleagues have shown that we can 'see' the bomb in the upper arm and survive. In other words, we can see the bomb, without any light reaching it. At play here is a new quantum principle: probability amplitudes must change when indistinguishable alternatives become distinguishable. In another form, this is known as the uncertainty principle, a good name as we will see. It is also the principle that permits quantum cryptography to be shielded from any eavesdropper.

Here is how Kwiat's scheme works. Set up the device in such a way that, when no bomb is present, photons emerge with certainty at the L-detector. In this case, the amplitudes for the histories RR and TT will cancel exactly, while those for RT and TR reinforce exactly. When we detect a photon at L the only thing we know for sure is that either RT or TR occurred, but each alternative is indistinguishable. If we also knew if the photon was reflected at the first beam-splitter we could distinguish these two alternatives. However, it is easy to see that any attempt to do that must render detection at L no longer certain. This is just a question of consistency. Suppose we put the bomb into the upper arm and it does not go off. Then we know for sure the photon took the lower arm. Now when it encounters the second beam-splitter it is as if we have a single photon approaching a single beam-splitter, and we already know that in such a case the photon must make a random choice to be reflected or transmitted. Thus if the bomb does not go off, there is an equal chance the photon will be detected at U as well as L. The non-explosion of the bomb has made two indistinguishable alternatives dis-

Figure 1.9 Kwiat's experimental scheme for demonstrating quantum seeing in the dark. The optical device is set up so that when the 'bomb' beam-splitter is not inserted, photons are never detected at the D-detector, that is, this output is always dark. However, when a beam-splitter is inserted that can direct some photons to a 'bomb' photon detector, the D-detector will sometimes record a photon even if the 'bomb' detector does not.

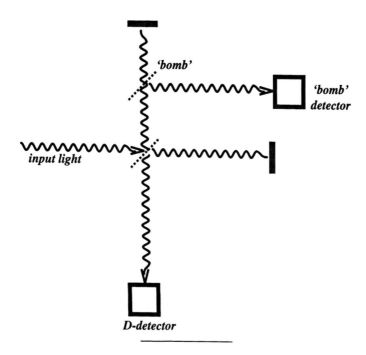

tinguishable. We know which way the photon went at the first beam-splitter and thus the possible detection events must change. No longer is detection at L certain and detection at U impossible, as it would be without the bomb. Now there is a 50 per cent chance that we will detect a photon at U. See Figure 1.9.

If we ever get a photon at U we know for sure the bomb was in the upper arm. Yet the bomb did not go off. If it did we would not detect photons anywhere, ever again. The bomb did not go off because no photons touched it, yet if we get a detection at U we know for sure the bomb is in the upper arm. We can 'see' the bomb without any light reaching it. Of course this is a rather dangerous game to play. You can only see the bomb in the dark 50 per cent of the time, for the other 50 per cent a photon takes the wrong turn at the first beam-splitter and your days as a quantum mechanic are over.

Quantum seeing in the dark reveals a new principle which modifies Feynman's rule. Recall what that rule said: if an event can happen in two indistinguishable ways, the probability amplitude of that event is calculated by summing the probability amplitudes for each of the indistinguishable ways the event can happen. In Kwiat's bomb detection scheme we must first set the device so that detection at L is certain. This is done by adjusting the path lengths in the upper and lower arm. Under such circumstances, we cannot know what the photon did at the first beam-splitter, as certain detection at L can happen in two indistinguishable ways, RT or TR. However, when the bomb is in the upper arm we can distinguish what happens at the first beam-splitter. We can know which path the photon took. If the bomb explodes, our lamenting colleagues will know that the photon took the upper arm. If the bomb does not explode, but we detect a photon at the U-detector, we know for sure the photon took the lower arm. The price we pay for certain knowledge of which path the photon took is to make detection at L completely uncertain. We cannot, in principle, ever set up a device in which detection at L is certain, yet know for sure which way the photon went. If we know the path, the detector is uncertain. If the detector is certain we cannot know the path. This is called the 'uncertainty principle', quite a good name I think, but really it is required if quantum mechanics is to be self-consistent. If we know which of a number of alternatives is realised, all future observations must be consistent with that fact. In the case of the bomb detector, if we know for sure the photon took the lower arm, it must behave at the last beam-splitter just as if it was encountering a beam-splitter for the first time. Only if the past history of the photon is in principle unknown, do we apply Feynman's rule and predict a result that is different from simply assuming that photons make random choices at beam-splitters. The uncertainty principle and Feynman's rule work together to ensure that quantum mechanics is both consistent and surprising.

It must be admitted that there is something decidedly spooky about quantum seeing in the dark. However, the experiment has been done—not with real bombs of course— and the results confirm quantum mechanics. To demonstrate the effect, Kwiat together with Weinfurter and Zeilinger and also Thomas Herzog, now at the University of Geneva, set up the device shown in Figure 1.9. The bomb has been replaced with another beam-splitter and an explosion is replaced with the rather less dangerous event of a photoelectric emission in a detector, called the bomb-detector in the figure. The optical arrangement is also slightly different, but nonetheless the horizontal and vertical paths can be adjusted so that, in the absence of the beam-splitter 'bomb', the photon is never detected at the D-detector. Introducing the extra beam-splitter into one arm changes this. In the experiment, the 'bomb'-detector registered a photon about half the time, but in 25 per cent of cases the D-detector. In such cases, the photon is not lost at the 'bomb' beam-splitter, yet because we know the detection took place at the D-detector we know for sure that the beam-splitter is there. The performance can be improved by making the central beam-splitter less and less

31

reflective, so that the chance of a photon taking the path with the 'bomb' is reduced. In this case, the probabilities for detection at the D-detector and detection at the 'bomb'-detector become equal. Half the measurements indicate the 'bomb' is there, but is not detonated.

Bits and qubits

All of recorded human knowledge can be reduced to one incredibly long string of ones and zeros which, to a casual observer, would look little different to the recorded results of a very long sequence of coin-tosses. The text of this book is represented inside my computer as just such a long string. Of course, we will need some kind of efficient code to represent the words and punctuation, the pictures and the formatting in such a simple form. Digital computers store numbers and information in terms of just these two symbols. The reason for this is that it is very easy to physically represent such a code in the electronic hardware of a computer. If a switch is ON it could code for a one and if it is OFF it could code for a zero. Of course we would not like an ON switch to suddenly and randomly turn to OFF, so in practice we need to set up our hardware to monitor and restore errant switches. With only two settings, such a correction scheme can be extremely robust, and this is one of the main reasons why computers use such a simple code. A code built out of just two symbols, 1 and 0 say, is referred to as a binary code.

We could have used a more complicated code if we had wished. Nature, in fact, codes biological information in a system of four symbols. In living systems, all the information required to build an organism and enable it to function is encoded in the sequence of base pairs aligned on the opposite strands of the DNA molecule. There are four chemically distinct base pairs, Adenine (A), Guanine (G), Cytosine (C) and Thymine (T). The last of these is replaced by Uracil (U) in molecules of RNA. A simple universal code of combinations of three base pairs is used by life to construct amino acids out of which are made proteins, the stuff of living things. A gene is the name given to a sequence of these four bases (not necessarily contiguous) on a strand of DNA that codes for some functional protein.

Information is a slippery concept that seems to have a great deal of subjectivity to it. One person's information is another person's noise. The information encoded on a compact disc recording of a heavy metal band may mean a lot to some people, but to me it is just noise. The large binary string that encodes this book in my computer would look little different to the results of a very long coin-toss, with a head represented as a one and a tail represented as a zero. It may sound surprising, but a quantitative theory of information can only be formulated in terms of randomness. This seminal insight was made by Claude Shannon, at Bell Laboratories, in 1948.

Shannon's idea is this: when we learn the result of a fair coin-toss, we gain one *bit* of information. In any experiment with two mutually exclusive outcomes, such as a coin-toss, we can code the results in binary form. That is to say, we can represent the two distinct outcomes by the numbers one or zero. When the result of a fair coin-toss is reported, information is gained. This elementary observation connects, somewhat paradoxically, random processes and the scientific measure of information formulated by Claude Shannon. The information acquired when one of two equally likely outcomes is realised is one bit of information. To store the result requires one binary digit of memory. Now we know that a quantum coin-toss can be quite different from a classical coin-toss. Does this mean that we need a new information theory to describe such experiments? How much information is acquired when the result of a quantum coin-toss is reported?

The behaviour of a single photon at a beam-splitter is like a fair coin-toss: it can be reflected or transmitted with equal probability. When we learn which possibility is realised we gain one bit of information. We could code reflection as a one and transmission as a zero. We could also decide to encode our bit of information in terms of the two mutually exclusive possibilites for photon polarisation. A one could encode a vertically polarised photon and zero a horizontally polarised photon. Quantum mechanics indicates that the world at heart is irreducibly random. At first sight this might suggest there is some connection to be made between information theory and quantum mechanics. But we immediately suspect a problem: the randomness of quantum mechanics is not like ordinary everyday randomness. It is governed at a deeper level by probability amplitudes and Feynman's rule. A large number of mathematicians and physicists are currently working on understanding the connection between information and the quantum theory, and the full picture is not yet in place. However, a number of important points are already clear. To begin with, if the quantum theory indicates the universe is irreducibly random, then the universe is an inexhaustible source of information, a bottomless reservoir of surprise!

Suppose we toss a single coin, but cover it so that the result is hidden. We know that it lies with head up or tail up, and for a fair coin we assume that each possibility is equally likely. When we look, and discover the coin is head up, we get one bit of information, and we have reduced uncertainty to certainty. If we cover the coin and look again a short time later it will still be head up. A photon at a beam-splitter, however, is different. We know there is no hidden variable of transmitivity (T) or reflectivity (R) that will determine the outcome of an experiment. So when we see a single photon is transmitted we cannot say, ah ha!, we have a T-type photon. If we send that photon through a second identical beam-splitter it is still just as likely to be reflected as transmitted. Unlike a coin, we cannot say that the observation has revealed some property of the *photon*. If we continue to send the photon through an endless sequence of beam-splitters, each encounter is totally random and each encounter will provide one bit of information.

34

Photon polarisation likewise cannot be regarded as a property of a photon, like the two distinguishable sides of a single coin. We can first select a particular set of polarised photons by filtering a light beam through a polarisation filter. Now we subject each photon to an encounter with a polarising beam-splitter (PBS). This is an optical device which redirects photons according to their polarisation. For example, all horizontal photons will be reflected and all vertical photons will be transmitted. What the polarising beam-splitter regards as horizontal is, of course, relative to the orientation of the device. If we rotate the PBS with respect to ourselves, its horizontal and vertical directions will rotate with respect to our definition of horizontal and vertical. If a photon is transmitted by the PBS we may be tempted to conclude it is a vertical photon. If we send that photon through another identical PBS, at the same orientation, it will of course be transmitted. Again this is the consistency required if physics is to be possible at all.

If we rotate the second PBS, we can no longer be certain if the photon will be reflected or transmitted. If it is transmitted, it is no longer polarised in the same direction that it left the first PBS. We can continue to send this photon through an endless string of PBS devices, all rotated by a slight amount with respect to the first. At each encounter, there are still two possible outcomes, transmitted or reflected. We can never say at any stage that this photon definitely has a preferred polarisation, fixed once and for all. Every PBS encounter is a new question, but to every question there is a single yes/no answer. We can ask a photon an infinite number of questions, but we will only ever get one of two possible answers. This is very different to a classical coin-toss, in which we reveal one of two hidden properties. In that case there is only one question.

A single photon has the capacity to provide an endless stream of information. This is a direct consequence of the quantum principle, that nature is an irreducible source of randomness. Classical physics, the physics of Newton and Einstein, is not like this. Uncertainty can always be totally removed if we have sufficient knowledge about a physical system and that knowledge can be gained by suitable experiments. In reality, however, the world is not like that. No amount of experimentation will ever remove the irreducible uncertainty of the quantum. That is one of the most difficult things to accept when one first encounters the quantum theory.

But there is more to it than that. It is not just that the world is irreducibly random. It is random in a very special way, reflected in the unusual rules for calculating odds from probability amplitudes. Can we really say that we gain one bit of information when we learn that a single photon is reflected or transmitted at a beam-splitter? Does this not misrepresent the highly unusual nature of quantum randomness and hide it behind the classical notion of a bit? True, we can code reflection and transmission in a binary form, but we know it is not just a coin-toss. We can see it is more than a coin toss by returning to the experiment in Figure 1.3. If a photon simply tossed a coin at every encounter with a beam-splitter, we would detect a photon at U or L with equal probability and changing the path length in the two arms would have no effect. But we know that we can change the detection probabilities for a single photon in this device simply by manipulating the path length. The quantum theory accounts for this in terms of probability amplitudes.

A quantum event, in which only one of two mutually exclusive outcomes is possible, is the elementary act of observation, upon which all knowledge of the physical world is built. The encounter of a single photon with a beam-splitter is an example. An elementary act of observation is at once like a coin-toss and not like a coin-toss. The ability to manipulate probability amplitudes is a new feature of the world not captured in a single coin-toss. It is true we do gain a single bit of information in an elementary act of observation, but there is more to it than that. To capture this difference a new word, *qubit*, was invented by Bill Schumaker at Keyon College, Ohio.

The physical state of a system, revealed in an elementary quantum observation, is not represented by a single bit, it is represented by a *qubit*. A qubit is the fundamental atom of uncertainty. Observations made upon a qubit can be as random as a coin-toss, or as certain as a mathematical truth. It all depends on the question, and we can ask an infinite number of questions. A qubit is infinity in a coin-toss. How can something as ordered and as complex be built on irreducible uncertainty? The answer is, build it out of qubits. To rewrite a well-known expression of the American physicist John Wheeler, quantum physics is all about getting It from Qubit.[7]

CHAPTER TWO

QUANTUM ENTANGLEMENT

Darwin's hidden variables

How did Darwin do it? In the teaming diversity of life on Earth, Darwin saw a new principle for the organisation of matter and the emergence of design. He was totally unaware of the gene or DNA. What ideas he did have on the principles of inheritance turned out to be wrong. Yet, despite his ignorance of the hidden variables of inheritance, he saw in the world around him abundant evidence for his new theory, and presented this evidence in an overwhelmingly convincing treatise titled *The Origin of Species* in the year 1859. How is it possible to know nothing about the molecular basis of evolution, or even of basic genetics, and yet still arrive at the correct theory? The answer is quite clear. The hidden variables of inheritance imprint their character on the relationships between all organisms. With minor change, these variables endure through generations, and are mixed like a multifold coin-toss in the game of life. The incessant experiment that is natural selection is constrained to produce results in which the correlations inherent in the genotype are written in the phenotype.

Darwin arrived at his theory, not by experiment, but by simply observing the outcomes of an experiment that had been running for billions of years. Every organism alive today is a data point in the theory of evolution. You cannot conclude anything by looking at a single datum, you must search for relationships in the data, you must look for *correlations.* All science is based on this method. The results of experiments, the results of observations, are not simply random facts thrown up by a gigantic cosmic roulette wheel. All observations are constrained by a fundamental, perhaps unknown, reality which forces relationships between observations. That is how Darwin did it. He did not need to know the fundamental principles of inheritance, they were written in the world for him to read.

Curiously, however, they were also written in a rather obscure scientific journal, *Transactions of the Brun Natural History Society.* Darwin, however, appears not to have read it. In 1865, Gregor Mendel, an Augustinian monk living in Bohemia, published the results of his experiments on peas. Mendel discovered that the principles of inheritance are based on the hidden variables we now call genes. The key methodological innovation that led Mendel to his great, but unheralded, discovery was statistical analysis of a modest kind. Mendel started counting traits in populations of peas.

If you are going to do statistics, then you must survey a large, even a very large population. If you are going to use probability reasoning you must repeat the experiment many times. In the play *Rosencrantz and Gildenstern are Dead*, Tom Stoppard begins with one of his characters throwing a coin over and over again and it always comes down heads. If it was thrown just once and came down heads we wouldn't even notice. If it came down heads twice in succession we might be mildly surprised. But if it came down heads in many successive trials we are rightfully astonished, just as Stoppard wants us to be. It is no ordinary world (or at least no ordinary coin) in which a tossed coin always comes down heads over and over again. This is a correlation of a very unnatural kind.

If the world around us was indeed a gigantic coin-toss, science would be impossible. Clearly, however, it is not entirely predictable either. All our hard won facts and theories about the world are sifted through a fine veil of uncertainty, randomness and error. The essence of the experimental method is to try to reduce these errors, to

39

mitigate the effects of uncontrollable factors so that the true relationship between observations may be discerned. Nonetheless, the relationship we seek may need to be extracted from the rough ore of experimental data by some very considerable statistical processing. Even something as simple as measuring the time taken for an apple and a pear to fall must be repeated so that small variations in the conditions of the experiment can be averaged out. Only then do we discover the remarkable fact that all objects accelerate at the same rate under gravity.

A great deal of randomness is evident in the facts of inheritance. Take sex, for example. The result is essentially a coin-toss, every time. If a couple has had six boys, the probability that their next child is a girl is no different to the probability that their first child was a girl, that is, close to 50 per cent. Of course a run of six boys is a little unusual, as would be a run of six heads in a succession of coin-tosses. We know that the probability of getting a boy or a girl is close to 50 per cent because in the population at large this is roughly what we find. Of course there is no one calling heads or tails to determine the sex of your child. The result arises from our genetic nature. Indeed, the fact that a survey of the population at large shows us that there is roughly an equal chance of getting a boy or a girl tells us something very strong about the principles of sexual genetics.

Other traits are somewhat harder to pin down. To begin with, they may depend on the random circumstances of your family tree. Each toss of the reproductive coin does not yield a result entirely random, but reflects the possibly erratic mating choices of your forebears. The history-dependent character of the odds with which various traits turn up is one of the reasons that people have such a hard time with probabilistic reasoning in genetics. Indeed even in mathematics, random events that are strongly dependent on past realisations are also rather hard to understand, and indeed often seem not to be random at all. Furthermore some traits do not have a simple yes/no character. There are shades of grey. Suppose, for example, we seek to determine the relative frequency with which the trait of genius occurs in a population. A smilar idea occurred to Darwin's cousin Francis Galton, who in 1869 published *Hereditary Genius*. The trouble is, genius is a rather vague concept to pin down. It is not quite so black or white as sex, not quite so binary. Mendel's great achievement was made by undertaking statistical analysis of traits which came in only two forms. In essence he sampled large populations of garden peas with a collection of simple yes/no questions.

Garden peas may not seem like a very interesting subject in which to search for the principles of inheritance. They don't play music, have never been recommended for their genius and on the whole don't display the huge multiplicity of traits that we regard ourselves as displaying. In fact, that is precisely the reason why peas are a good subject in which to study inheritance. They exhibit a small number of easily identifiable traits, such as height and flower colour. Mendel used traits in peas which had a yes/no character. As we shall see throughout this book, forcing nature to answer a sequence of yes/no questions can provide a powerful route to the underlying laws. Human traits such as mathematical ability, disposition to crime and so on, do not have such a simple representation. There are shades of grey that are not easily captured in the answers to yes/no questions. By looking for the presence or absence of simple traits, Mendel was able to clearly display the logical pattern of inheritance in the frequency by which certain traits appeared in populations.

We are going to be doing a lot of reasoning with probability, so let us take a closer look at what Mendel did. It will provide as good an introduction to reasoning with probability as any. Mendel began with two pure breeding lines of peas, one line always had tall offspring and the other had short offspring. There were never any intermediate heights. If your parents, grandparents, great grandparents . . . were tall, you were tall. Likewise if all your family tree consisted of short peas. I will call these two types of pure breeding lines the P-generation. Now the interesting thing about peas is that they can be self-fertilised, so that a single plant is both father and mother to the offspring. This trick simply ensures that the 'parents' are identical. Mendel now crossed a short pea from the P-generation with a tall pea from the P-generation, and the resulting seeds (and there are, fortunately, very many of them) are planted out. I will call this bunch of plants the X_1-generation (X for cross). The results: every seed grew into a tall pea. The probability of getting a tall pea from this combination of parents is exactly one. It looks like whatever determined a short plant has vanished from the X_1-generation of peas. Such is not the case.

An interesting result obtains when each of these tall plants is self-pollinated. This means that each tall plant is crossed with an identical plant. Now the many seeds of the union, called the X_2-generation, are planted out and we find some short plants hiding among their tall siblings. Shortness is still present in the X_2 line of peas. This illustrates the most important principle of genetics, that is, genes are not diluted or destroyed in breeding, but are carried on through populations in some particulate measure.

Mendel then counted the number of tall and short plants in the X_2-generation. In one particular cross he obtained 1064 plants, of which 787 were tall and 277 were short. This is a good population size on which to do some statistics. Suppose we ask for the probability that cross breeding two tall plants from the X_1-generation will lead to a tall plant in the X_2-generation. Of course we cannot know for sure if it will be tall or short. From Mendel's data we might estimate that this probability is 787/1064, or roughly 75 per cent. Likewise the chance that a cross of two tall plants in the X_1-generation giving a short plant in the X_2-generation is about 25 per cent. If we assume Mendel's population was typical, and its reasonably large size would support this, we can assign a probability of 3/4 for tall and 1/4 for short.

The explanation for Mendel's results hinges on postulating hidden variables which determine the height of the plant. Mendel called these 'factors'. We will call them genes. How these hidden variables combine during pollination will determine the height of the plant. We now need a little genetics. Firstly the hidden variables, genes, reside in the germ cells of the plant (gametes) in a single copy. However, in the rest of the plant's cells they are in two copies. For example, a pure breeding, P_1-generation, tall plant has a single tall gene, T, in the germ cells, but two tall genes TT in all the other cells. If we self-pollinate these plants, only one member of the pair goes into the germ cell. However, as both are tall (TT), it is certain that the germ cell will have a T-gene. This is what causes the P_1-generation to always produce offspring that have the same height, that is, tall. A similar result holds for the short plants in the P_1-generation. Designate the gene that determines shortness by t. Then all short P_1 plants have a tt combination in all cells except the germ cells in which there is a single t.

Now when we cross a tall and a short plant from the P_1-generation, we throw together a T variable and a t variable. These end up in germ cells entirely at random. It is like tossing a coin with one side labelled T, the other t. One offspring may end up with a T, another with a t. However, the rest of the plant gets two copies, Tt. You might have thought that this would lead to a medium height plant. However, Mendel's experiment indicates that all plants in the X_1-generation are tall. Clearly there is more to genetics than statistics. In this case, the gene for tallness is said to be dominant. If all the plant cells get a Tt combination, the plant will be tall as the T gene is dominant. We don't need to worry about how this might come about, we simply take it on board as a fact of life for these kinds of hidden variables. This an important lesson. Even though hidden variables are involved in determining how the plants inherit height, other features come into play when hidden variables interact.

The germ cells of the X_1-generation have only a single copy of the gene, but what they get is random. A particular germ cell could be either a t or a T, with the same probability for each, that is, 50 per cent. Let us move onto the X_2-generation by self-crossing one of the X_1 plants. We do not know for sure just what germ cells are going to combine. It could be either a t or a T. This is like tossing two coins with a T-side and a t-side on each. The possible results are TT, tt, Tt and tT. We can apply Laplace's rule of insufficient reason to assign the probability that a plant in the X_2-generation will have any one of these genes. So the chance of getting say TT is 25 per cent. But a Tt is indistinguishable from a tT. Following Bayes' rule we thus add the probabilities for the two ways to get an oddly matched pair. Thus the chance of getting an odd pair is just 50 per cent. What is the chance that the plant in the X_2-generation will be tall? We now need to use the extra little fact about plant genetics that we learnt in the previous paragraph. As T is dominant there are three indistinguishable ways a plant can be tall. It can have a TT gene or an odd combination (Tt or tT). Again using Bayes' rule we add the chances for each way, to get 25 per cent + 50 per cent, that is, 75 per cent. In a large population of offspring from a self-cross of the X_1-generation plants we will find roughly 75 per cent tall plants in the X_2-generation.

Mendel was able to arrive at an understanding of the principles of inheritance using simple probability rules. These rules enabled him to work back from the observed frequencies of inherited characteristics to discover the hidden variables that determine, incompletely, the traits passed on from one generation to the next. These variables are passed from one generation to the next in random combinations, but are not destroyed or diluted. They do of course interact in subtle ways, as the discovery of dominance shows. The kind of reasoning used by Mendel, based squarely on simple rules for combining probabilities, is used everyday by natural scientists, social scientists, economists and possibly even hopeful speculators at horse races and casinos (although rather inadequately I suspect). The rules of classical probability are one of the fundamental mathematical tools that enable the scientific endeavour to proceed. It thus comes as quite a surprise to learn that early experiments in quantum physics in the first few decades of this century gave results that simply could not be explained by classical probability theory operating on physical variables, hidden or otherwise.

One of the most surprising of all such experiments was, in fact, not performed in any definitive way until 1982. This was the Aspect experiment performed by Alain Aspect and co-workers in Paris. Like all good experiments it was done to resolve a puzzle, first posed by Albert Einstein, Boris Podolsky and Nathan Rosen in 1935. Their work, now known as the EPR paper, was the genesis of the Aspect experiment and the genesis of the most active area of contemporary physics and the subject of this book. What EPR drew our attention to was the very puzzling business of *quantum entanglement*.

A bolt from the blue

The central lesson of the quantum theory is a deep and surprising fact about our world: it is irreducibly random. Why irreducible? Ordinary everyday randomness, such as the toss of a coin, is supposed to result from our ignorance of all the factors controlling an experiment. The ignorance does not result from laziness. Often there are just too many factors that would need to be known to reduce the randomness to certainty. A classical physicist, one not in receipt of the quantum principle, might claim that the behaviour of all the water molecules in a steam-filled cylinder could be predicted with certainty if only we knew the initial position and velocity of each and every molecule. Needless to say this is not a very practical hope and indeed not even a useful research goal. We have no trouble predicting the behaviour of steam pistons with a more coarsely grained description known as thermodynamics. Nonetheless a classical physicist would continue to believe that the analysis of thermodynamics could be taken further.

Indeed it was taken further by, among others, Ludwig Boltzmann and Albert Einstein. They showed how to supplement classical Newtonian physics with statistical ideas to provide greater insight into the behaviour of large numbers of particles. Einstein's particular contribution was an explanation of the Brownian motion, the apparent erratic motion of small pollen grains suspended in water, first observed by Robert Brown in 1828. Brown assumed that the motion was due to something in the pollen grains themselves. However, even lifeless particles exhibited the motion, incuding fragments chipped from the Sphinx! The correct explanation is quite simple. The pollen grains are being constantly bombarded by water molecules. Einstein's analysis of this phenomenon, published in 1905, was in fact one of the clinching arguments for the atomic hypothesis. Despite the recourse to statistics, however, a classical physicist such as Einstein would never doubt that the particles of a gas exist and possess a position and velocity at every instant. In the face of the quantum theory, however, this faith in the reducibility of randomness to certainty cannot be maintained. Needless to say, Einstein saw this as evidence that the quantum theory was not quite the full story.

Einstein was a truly creative and revolutionary thinker. He was himself responsible for one of the central features of the quantum theory of light, the photon, a small indivisible quantum of energy for light. However, he was also a classical physicist. For such a person, the claim that our world is irreducibly random is not going to go by without a challenge. The resulting clash between the classical ideal and the quantum principle became embodied in a clash of two intellectual titans: Albert Einstein versus Neils Bohr, another of the founding fathers of the quantum principle. At stake was our conception of reality itself.

The first round of this engagement took place on the afternoon of 27 October 1927. On the morning of that day, the leading atomic physicists had gathered at the Hotel Metropole in Brussels for the Fifth Solvay Conference. The purpose of the meeting was to try and reach an agreed understanding of a new theory of the atom that had been advanced in two different mathematical forms by Werner Heisenberg and Erwin Schrödinger. The proper interpretation of either form was, however, far from obvious. Neils Bohr and Albert Einstein also attended the meeting. Bohr had a new interpretation of the theory that was in complete conflict with the classical ideal that had served physics so well for over 300 years. Bohr presented his views to the immediate, and entirely expected, objections of Einstein. Thus began the debate that would shift the classical paradigm forever.

I am not going to recount in detail the debate between Einstein and Bohr on the interpretation of the quantum theory. This has been detailed in many books and articles.[1] Most physicists (including me) accept that Bohr won the debate, although like most physicists I am hard pressed to put in words just how it was done. The discussion was based around a number of ingenious proposals by Einstein for ideal experiments which attempted to show in one way or another that the quantum theory, though a statistical theory, could not be complete and that the analysis would need to be carried further. Initially these idealised experiments centred on the two uncertainty principles in quantum theory which appeared to constrain, in principle, the classical ideal of reducing a statistical theory to a non-statistical theory. It is undeniable that Bohr's explanation of these experiments was correct and established the interpretive principles on which quantum theory came to be based. However, Bohr's response to the last salvo fired by Einstein in the debate cannot be described in quite such definitive terms. In 1935, the paper by Albert Einstein, Boris Podolsky and Nathan Rosen, now referred to as the 'EPR paper', became known to Bohr. As one of his assistants at the time reports, 'This onslaught came down on us as a bolt from the blue. Its effect on Bohr was remarkable.' Bohr saw in the EPR paper the true consequences of the quantum theory for many particle systems, systems that would soon be called entangled systems by Erwin Schrödinger, one of the founding fathers of the quantum theory, for it is in entangled systems that the truly revolutionary character of the quantum view of reality becomes most apparent.

In the EPR paper, the quantum theory was applied to a system of two particles, which interacted with each other at some initial time, and then moved off in opposite directions. After some time, and after they have moved a very great distance apart, the particles enter a measuring device to determine some classical property of each of them independently. For example, the momentum could be measured on each particle. Or perhaps the position of one could be measured and the momentum of the other measured. The important point is this: because the particles had interacted in a special way in the past, there will be strong correlations between the results of measurements on each of the pair. For example, if one particle is found to have a particular momentum, the other of the pair will have exactly the equal and opposite value for momentum. Because of this perfect correlation we only need measure the momentum on one particle and the result can be applied to the other particle. In other words, we can predict with certainty the result of a momentum measurement on the second particle *given* the results of a measurement on the first particle.

None of this sounds very surprising. Such strong correlations are a common feature of the world. If I put a blue and red marble in a box and remove a marble without looking, transport it to the farthest reaches of the galaxy and then look to see that it is blue, I can predict with certainty that the other marble back on Earth is red. We have to dig a bit deeper to see why Bohr or anyone should bother with such a trivial point. The marble analogy is misleading in an important way. The marbles are only distinguished by their colour: we can only ever measure colour. However, for the particles in the EPR paper we can do more. We can measure momentum or we can measure position. Now it turns out that the positions of the particle are also tightly correlated, as you might expect if their momenta are strongly correlated. So instead of measuring momentum on one particle we could equally well have measured its position. Then we can predict with certainty the position of the other of the pair without interacting with it in any way. What is so surprising about that?

One of the central claims of the quantum theory is that there are *no* physical states whatsoever in which a particle can simultaneously have perfectly precise values for momentum and position, a result called the uncertainty principle. But the EPR argument seems to suggest that for one of the particles, the one we don't measure, we can predict with certainty either its momentum or its position, without interacting with it in any way. If it is in any state at all, it must have had precise values for position and momentum before we measured its twin on the other side of the galaxy. What happened to the uncertainty principle? The EPR argument is absolutely watertight given its initial premises. How could Bohr, or anyone who understands the quantum theory, get around this one?

The conventional quantum answer lies in the notion of a *conditional state*. The state of a particle is a complete description of the preparation and circumstances of a system that enable us to determine the results of measurements made upon that system. The uncertainty principle refers to a single state of a particle. It implies that if we prepare a whole bunch of particles each in the *same state*, a state for which the position is precise, but we choose to measure the momentum instead, we will get a large spread in our data. Conversely if we prepare a bunch of particles each in the same state, for which the momentum is precise, but choose to measure position instead, we will get a large spread of results. However, this is not at all what happens in the EPR proposal, at least for someone arguing entirely along quantum lines. The problem is this. If we choose to measure the position of one particle and get a result, the conditional state of its twin must now change to reflect our increased knowledge and the previously understood correlation between the two particles. This conditional state is clearly one in which the position is precise. However, if we choose to measure the momentum of the first particle, the conditional state of the twin *is different*. It is now a state in which the momentum, not the position, is precise. We cannot combine predictions made for systems in *different states* and claim a violation of the uncertainty principle. This is precisely the point that Bohr made, somewhat obliquely it must be admitted, in his response to Einstein:

> The criterion of physical reality proposed by [EPR] contains an ambiguity as regards the meaning of the expression 'without in any way disturbing the system'. Of course there is in a case like that just considered no question of a mechanical disturbance of the system under investigation during the last critical stage of the measuring procedure. But even at this stage there is essentially the question of *an influence on the very conditions which define the possible types of predictions regarding the future behaviour of the system*. Since these conditions constitute an inherent element of the description of any phenomenon to which the term 'physical reality' can be properly attached, we see that the argumentation of the mentioned authors does not justify their conclusion . . .[2]

Einstein was a smart guy, so how could he have missed this obvious point? The fact is he didn't. He and his co-workers were not working with the quantum meaning of *state*.

However, neither Bohr nor Einstein had quite recognised all the implications of the entangled states discussed in the EPR paper, implications it has taken decades to draw out. Schrödinger saw a little further in his famous 'cat' paper which he was inspired to write upon reading the EPR paper. I will not discuss Schrödinger's ideas here, but will postpone his notorious cat treatment until later. Many years later, David Bohm, working in Princeton, rephrased the EPR argument in terms of a simpler system for which the possible results for measurements are finite, rather than the infinite continuum that obtains for position and momentum. The real breakthrough came in 1963 when John Bell, an Irish physicist on sabbatical leave in Stanford from CERN (the European high-energy physics laboratory), wrote two papers drawing out the revolutionary consequences of the entangled quantum state. However, to see what Bell saw we must prepare ourselves. To be surprised by Bell's truly surprising discovery we had better become familiar with the . . . well, with the familiar!

The puzzling business of quantum entanglement

The citizens of a large American city, lets call it Bellsville, tiring of consumer surveys, decided to act. For decades, information on all sorts of things had been collected. Considerable ingenuity had been exercised to extract information efficiently. Sometimes it was as simple as a direct phone survey. Other methods were more devious. When it became common to have to fill out a survey form in order to buy a cup of coffee, the citizens decided that enough was enough. A referendum was held to determine support for proposition 666: the *Electronic Privacy Regulation* (EPR). The new regulation was based on the concept of 'bit restriction', devised by Professor M. Jackson, an expert on privacy protection at the Central Intelligence Agency. In effect the law stated that no more than a single bit of information could be held for any individual. This meant that in any survey, the answer to no more than a single yes/no question could be recorded for any single person.

Advertising corporations, however, are staffed by highly intelligent and creative folk. One such corporation, Aspect Inc., came up with a way to beat the law. They decided to only survey identical twins. Their reasoning was this. If we assume identical twins share a common genetic code, then we can ask one question of one twin and a different question of the other twin, and the answers will apply to both twins.

Aspect Inc. decide to conduct a survey to demonstrate

the strong genetic correlation of identical twins. It is decided that three traits will be recorded for each member of a pair of identical twins drawn at random from a large sample of Americans living in Bellsville. The three traits are: sex (S), height (H) and eye colour (C). In the case of sex it is decided to record a minus sign (–) if a twin is male and a plus sign (+) if a twin is female. In the case of height a plus (+) is recorded if a twin is short (less than 1.8 metres) and a minus (–) if a twin is tall (more than 1.8 metres). Finally, in the case of eye colour a plus (+) is recorded if a twin has blue eyes and a minus (–) if a twin does not have blue eyes.

Being good corporate citizens, Aspect Inc. are quite happy to go along with the EPR law. They are thus going to record the answer to a single question for any given individual. They can of course ask different questions for each of a pair of identical twins, which is precisely the way in which they hope to get around the EPR. However, in order not to bias the sample, they decide to ask questions of each twin at random. The data is recorded in a list with two parts, one for twin A and the other for twin B. A small extract from the data is shown in Table ·2.1.

Examination of the entire data set shows a number of interesting features. If we look down a single column for either twin, say sex, we find approximately an equal number of pluses and minuses. This indicates that there is roughly a 50 per cent chance that a twin will be male. A similar result is obtained for the other traits. Thus there is a 50 per cent chance that a twin will have blue eyes. An obvious result is obtained when we compare the result for the same trait on each twin of a pair: both twins exhibit the same trait. Thus if twin A has blue eyes, twin B has blue eyes; if twin A is tall, twin B is tall; and if twin A is male, twin B is also male. This is an example of a perfect correlation. Such a correlation indicates that we only need determine the trait of one twin of a pair and we know for sure the corresponding trait of the other. When we determine the trait of one of the twins we have all the information about

Table 2.1 Extract from a survey of genetic correlation of identical twins for sex, height and eye colour

TWIN A			TWIN B		
SEX	HEIGHT	COLOUR	SEX	HEIGHT	COLOUR
+			+		
	−		+		
	+				+
−				+	−
+		−			
	−		+		
				−	

A small extract from the data resulting from a survey of twins. Each twin is asked one of a set of three questions. The three questions relate to sex, height or eye colour. Only a single yes (+) or no (−) answer is recorded. Further, only a single yes/no result can be recorded for each individual in compliance with the Electronic Privacy Regulation (EPR). The question asked of any particular individual is random and thus it is not always the case that the same question is asked of each member of a given pair. The data for each member of a pair is recorded in each row. For ease of compilation one twin of a pair is designated A and the other is designated B. Note that when the same question is put to a pair of identical twins, the answers are the same. This is a perfect correlation.

that trait for the pair. We gain no further information by also determining that same trait on the other twin.

Now the whole point of conducting the survey was to get more than a single bit of information about a given individual. To do this we need to look for entries where twin A was asked a different question from twin B. Because of the perfect correlation between twins discussed in the previous paragraph, any information gained on one member of a pair can be applied directly to the other. However, there is no way to get perfect information on all three traits simultaneously. At least one trait always remains unknown.

This will turn out to be very important when we come to the quantum analogue of this survey.

The chief statistician of Aspect Inc. has an amateur interest in genetics and wonders if there might not be some interesting hidden correlations between eye colour and height. Perhaps, for example, if we know the height of twin A we also have some information about the eye colour of twin B. Can we find evidence for such *cross correlations* in the data? For example, we might ask if a twin has blue eyes are they also likely to be male? Of course the EPR (Electronic Privacy Regulation) ensures that information on both these two traits may not be known for any individual simultaneously. However, for twins there is an easy way to do this precisely because the traits of each twin in a pair are highly correlated.

We extract from our data the results for different pairs of traits. For example, we could ask for the sex of twin A and the height of the corresponding twin B. Let us call this a SH survey. So as not to bias the survey let us also ask for HS whenever we ask for SH. In fact, let us extract all the data for which different traits are recorded for each of a pair of twins. There are six possible pairs of different traits: SH, HS, SC, CS, HC, CH. The data is recorded in a list like Table 2.2. In how many cases do we get the same result for each query? The result for this population turns out to be very nearly 50 per cent. That is to say, in roughly 50 per cent of cases when a different trait is examined for each twin of a pair the results are the same, that is both records have a + or both records have a −.

This result is exactly what would be expected if all possible combinations of the three traits were determined by a set of hidden Darwinian variables, genes, with all genes being equally likely to turn up. We all know that such things as sex, height and eye colour are determined by our genetic heritage. In some sense, what we see in each twin is a reflection of their shared genetic heritage. Let us assume for simplicity there are specific genes which determine whether a + or − results for each trait. As there are three traits (S,

54

Table 2.2 Extract from a survey of sex, height and eye colour in identical twins

SAMPLE	SEX		HEIGHT		COLOUR	
	A	B	A	B	A	B
1	+	+	−	−	+	+
2	+	+	+	+	−	−
3	−	−	−	−	+	+
4	+	+	−	−	−	−
5	−	−	+	+	−	−
6	−	−	+	+	+	+

A sample extract from the data record for a survey of three traits, sex, height and eye colour in a large group of identical twins. Each twin is distinguished by a label A or B. The traits are recorded as simple yes (+) or no (−) results. For example, is an individual female? If the answer is yes, record a +, otherwise record a −. Does an individual have blue eyes? Yes is +, no is −. Is an individual taller than 1.8 metres? Yes is −, no is +. Note that there is a perfect correlation between twins for any trait.

H, C) and two possible results for each (+/−) there must be eight different genes, as depicted in Table 2.3.

Each pair in a set of twins will have exactly the same gene, but otherwise we assume that each of the eight genes is equally likely. It is now easy to account for the observed data on the basis of simple counting arguments. First, if we query only the sex (S) for a particular twin we see that 50 per cent of the genes will give a −, thus we expect to see 50 per cent male twins and 50 per cent female twins. Likewise for the other traits. If we query the same trait on each twin in a pair we must get the same result, as each twin carries the same gene. Finally, if we query different traits on each twin of a pair we must get the same result 50 per cent of the time. For example, if we query SH there are two genes which give a ++ and two genes which give a −− , thus we expect that, for a pair of twins chosen at random, roughly 50 per cent will give the same results

Table 2.3 The eight gene set combinations of sex, height and eye colour

S	H	C
+	+	+
−	+	+
+	−	+
+	+	−
−	−	+
+	−	−
−	+	−
−	−	−

A summary of the eight sets of genes, the 'Darwinian hidden variables' that determine the sex (S), height (H), or eye colour (C) of each person. In the case of identical twins, each member of the pair has exactly the same gene set. For simplicity, each gene is specified in terms of a single yes/no result in the phenotype of the individual. For example, height is simply a + for short and a − for tall. Each individual in a pair of identical twins will get the same gene. However different pairs of twins draw any one of the eight genes at random.

if S is recorded on twin A and H is recorded on twin B. The same is true for all the other different queries.

I am implicitly using two simple rules for dealing with chance and randomness: Laplace's rule of insufficient reason and Bayes' rule. There are eight genes which completely determine the three traits of interest for each twin. If all are equally likely, then by Laplace's rule each twin has a 1/8 chance of holding any particular gene. What is the chance that a twin will have the gene (++?), that is, + for sex and + for height, but unknown for eye colour? To answer this we use Bayes' rule. There are two ways that this gene can occur, each distinguished by whatever the entry is for eye colour. Bayes' rule says that in this case we simply add the probability for each way, considered separately. Thus the probability to get (++?) is just $\frac{1}{8} + \frac{1}{8}$, that is $\frac{1}{4}$. A similar reasoning gives the probability to get $(--?)$ as $\frac{1}{4}$.

Now if we ask for the probability that a gene with the same code for sex and height can occur, we note that this can occur in two ways as well, either $(++?)$ or $(--?)$. So according to Bayes' rule we just add the probabilities for each way, considered separately. Thus the probability that a gene codes for the same result for sex and eye colour is $\frac{1}{4} + \frac{1}{4}$, that is, 50 per cent. Thus if we query different traits on each twin of a pair we must get the same result 50 per cent of the time.

Now get ready for the big surprise. If we lived in a world where the correlations between twins were determined, not by hidden Darwinian variables, but by quantum entanglement we would find that the number of cases in which the same result was recorded, when each twin of a pair was asked a different question, is very much less than 50 per cent. In fact, it could be as low as 25 per cent, *and at the same time still give a perfect correlation when the same question is asked of each of the pair.* How could such a result possibly obtain? The reasoning that led to the result 50 per cent was very elementary arithmetic supplemented by Laplace's rule of insufficient reason and Bayes' rule. Neither of these rules looks unreasonable. What can possibly go wrong with such a simple calculation? How could data not obey such simple rules of addition and multiplication? The answer of course takes us right to the core of the quantum theory.

The science (fiction) of quantum genetics

In the year 2005, a joint NASA–European Space Agency mission to Europa, a moon of Jupiter, confirmed that life did indeed exist in a sea of water beneath the frozen ice surface of the planet. At first sight, life on Europa looked rather on the dull side—small, single-celled bacterial organisms that appeared to have no variation. Subsequent analysis revealed, however, some surprising features and a distinctly puzzling life cycle. Every few Europa days, a pair of bacteria would coalesce and begin to swell to many times their original size. This larger organism would then attach itself to a source of ions of an unstable isotope, which was plentiful near volcanic vents at the bottom of the ocean. In a short period this parent organism would then produce millions of pairs of baby bacteria which would be quickly separated by the turbulent hot water near the vents. The bacteria were always produced in pairs, and planetary biologists became interested in whether or not the genetic codes for each pair of a twin were in fact identical.

Alas, every attempt to isolate the genetic material of the organisms failed, and it quickly became apparent that whatever was the molecular basis of the genetic code for these organisms it was highly unstable and easily destroyed. Following in the tradition of a century of Earth biology, the planetary biologists decided to map the genotype of each pair of twins by studying the phenotype, that is, instead of reading the information coding for the organism they would try to read the product of the code in the nature of the organism itself.

One of the most important facts about my imaginary Europa is the intensity of its magnetic field. Biologists conjectured that as life is a great opportunist, it was very likely to have evolved in such a way as to exploit the strong magnetic fields for a purpose. Indeed it soon became apparent that each little organism behaved like a small magnet, and it was supposed that this enabled the organism to orientate itself in the oceans of Europa. After a great deal of experimentation it was established that the built-in magnet of an organism was indeed subject to some variation and was very likely the key trait involved in inheritance and selection of this organism.

Recall that when these bacteria reproduce they always produce twins. One day, a physicist who happened to be passing though the Laboratory of Planetary Biology, recalling something he had once read about experiments back in the 1990s, suggested a simple experiment to measure

Figure 2.1 A device for measuring the direction of the magnet within each organism.

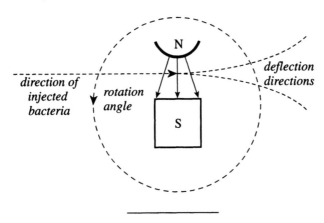

the magnetic direction of each of the twins. The north and south poles of two very differently shaped magnets were placed opposite each other (see Figure 2.1). The result is a magnetic strength that varies a lot from the north pole of the bottom magnet to the south pole of the top magnet. The two magnets are mounted on a platform that enables them to be rotated in the plane and a single organism can be injected through the device one at a time.

An artificial incubator for organism reproduction was established and as each pair of twins zoomed out of the parent body, they were separated and passed through the magnetic apparatus. A strong correlation was found. Any particular organism would be deflected up or down apparently at random, no matter what orientation angle of the apparatus is set. However, if one twin is deflected up the other is *always* deflected up and if one twin is deflected down the other is *always* deflected down. Of course no-one was too surprised at this. This result appeared to confirm what planetary biologists had suspected, a pair of twins is genetically identical. So, despite the considerable variation in the deflection from one pair of twins to the next, each member of a pair was very strongly correlated.

The next experiment sought to take each twin and pass it through another magnetic analyser, this time oriented differently in the plane. Alas, no result was obtained. It was soon realised that passing one of these little organisms through a magnetic analyser so deranged the internal arrangements of the beast that it died a few milliseconds after passing through a single device. It looked like only a single shot experiment would be possible on each individual. These organisms have a kind of built-in EPR, so that no more than a single bit of information, a single answer to a yes/no question, could be obtained. One could determine whether an individual is deflected up or down (a single bit of information), but the very act of getting that information removed any possibility of further measurements on the organism.

The planetary biologists studying these creatures then realised that they could do something similar to the survey undertaken by Aspect Inc. to get more than a single bit of information. Each individual was found over and over again to be perfectly correlated with its twin, and this of course indicated that each pair of twins had an identical genetic make-up that predetermined which way they would be deflected in the magnetic device. So, we can measure deflection of one individual with the magnet set in one direction and then subject its twin to a measurement with a magnet set in a different direction. As each twin is completely identical as far as these measurements are concerned, the information so obtained would be just as good as making repeated measurements on a single individual.

It was decided to use just three different orientations of the magnets: one setting at zero degrees, another setting at 120 degrees and the third setting at 240 degrees. However, now one twin could pass through a magnet set differently to its sibling. After many experiments the data were analysed. Then came the big shock. The results were deeply disturbing, throwing into doubt the foundations of probability used for over a century of experimentation.

What was the result? The number of twins which gave the same deflection (that is, both up or both down) for magnets set differently for each of the pair, turned out to be about 25 per cent of the total number of twins used in the experiment. Why so shocking? Recall the survey undertaken on Earth twins in which three different properties were measured. The result was that 50 per cent of all cases analysed had the same response where each member of a pair was examined for different traits: a result that was easily explained by Darwin's hidden variables, the gene, supplemented with the probability rules of Laplace and Bayes.

It is worth repeating the explanation again, this time for the trait of magnetic deflection. Three different settings of the magnets were used. In each case an organism is either deflected up or down apparently at random. The trait we are searching for is the predilection for deflection up or down with respect to a given orientation of the magnets. If for one setting, say zero degrees, an organism is deflected up we say that organism is + for that trait. If, however, it is deflected down we say that organism is − for that trait. As we use three different orientations, we can look for three traits. We will call these A, B and C. So, if trait A is +, we know that an individual was deflected up and its twin was also + for that trait.

Now following in the footsteps of earthbound genetics, we would assume that the traits exhibited are predetermined by the hidden genetic variables of the organism. There must be eight such genes corresponding to the eight ways of arranging +/− in each of the three slots A, B and C. To explain the perfect correlation when the same trait is measured on each twin we need only assume that each twin has exactly the same gene drawn at random from the set of eight possible genes.

We can only measure a single trait, A, B, or C, on each individual as the measurement destroys the organism.

61

But there is a way to get more than a single bit of information from a given individual. We can measure two traits simultaneously on the same genetic constitution simply by measuring a different trait on each twin. Two different measurements is the best we can do as by definition a twin means only two siblings. No matter what we do, at least one trait will remain unknown for a given pair of twins. If we measure A on one twin and C on the other we are completely ignorant of the outcome of trait B.

How many ways can we get the same result when a different trait is measured on each of a pair of twins? This is a repeat of the question asked by Aspect Inc. to explain the result of their survey of Earth twins. Here it is again to save you from looking back. There are eight genes, each of which is random, so the probability of getting any particular gene in a pair of twins is 1/8. This is Laplace's rule of insufficient reason. Suppose we look for the cases where the same result occurs when trait A is measured on one twin and trait B is measured on the other. Looking down Table 2.3 we see that there are two cases in which a ++ is recorded, and two cases in which a -- is recorded. Each of these differs in the unmeasured trait C, which could be either + or -. As we don't know what trait C is we simply add these two cases together to get a probability of 1/4 for ++ and a probability of 1/4 for --. If we now ask for the probability of getting the same result for AB (that is, ++ or --) we simply add 1/4 and 1/4 to give 1/2 or 50 per cent. Simple enough? Alas it does not work for these particular bugs! Only 25 per cent give the same result when two different traits are measured. If we insist on an explanation in terms of genes, strongly indicated by the correlations observed, the evidence points to a failure of classical probability. However, there is a new set of rules for dealing with probability, a quantum probability calculus, which can explain the results. The key to understanding why classical probability fails is the fact that for a pair of twins we can only ever know two traits simultaneously.

The explanation hinges on two new rules for dealing

with chance and combining probabilities. Laplace's rule tells us how to assign probabilities when we have almost no information to go on. It says that, if we know no better, assume all outcomes are equally likely. If we have eight possible genes we can assume that they will occur each with a probability of 1/8. Bayes' rule for ordinary probability says that if an event can happen in two ways, each of which is indistinguishable, we add the probabilities for each way, considered separately. So if we toss two coins the chance that they will come down odd (head/tail or tail/head) is just 1/4 + 1/4 or 50 per cent. In the case of Earth twins, when we choose two different traits to measure on each twin, say SH or HS, the third trait, C, remains unknown. Thus there are always two ways an outcome from the SH or HS measurement can occur corresponding to the two unknown outcomes (+/–) for C. Likewise for the organisms. We can only ever measure two different magnetic traits for each twin. The third always remains unknown. (You are probably thinking, well why not use triplets. Good idea. We will come to that. It is the subject of the GHZ scheme discussed in the next chapter.)

Now the first new rule is that probabilities are not assigned by Laplace's rule, but instead are calculated from a more fundamental object called a probability amplitude. To get a probability amplitude in the first place is not a simple matter—that is what you learn how to do in a quantum mechanics course. But once you get a probability amplitude the probability is easy to get, as we have seen. The next rule tells us how to combine probability amplitudes. It replaces Bayes' rule by a simple change: don't add probabilities, add probability amplitudes. As in Chapter 1, I will refer to this as Feynman's rule, after the famous Caltech physicist Richard Feynman.

The key point is that probability amplitudes, unlike probabilities, are not just simple positive numbers. Probability amplitudes can be negative numbers as well. When we start adding positive and negative numbers together of course there is a good chance that the final answer is going

63

to be negative as well. That is no problem as Feynman's rule says that to get the probability we have to square the answer. Now the square of any number is always positive and the probability comes out quite sensibly as a positive number. However, if we add a positive number and a negative number we may simply get zero, the square of which is still zero. So distinct probability amplitudes can be combined to give a probability of zero, which describes an event that never occurs.

Suppose now we have a pair of twin bugs. One of the twins is passed through a magnet oriented in direction A and the other is passed through a magnet oriented in direction B. How either of these bugs would behave if it passed through a magnet along the C-direction is forever unknown in this particular experiment. That is the key. No matter which way a bug is deflected in either magnet, up or down, it could do so without us knowing which of the two ways it might have been deflected in the C-direction. If one twin is deflected up in direction A and the other is deflected up in direction B (a result we record as ++), there are two ways this could have happened. Each way corresponds to the unknown outcome of a measurement in the third direction. Each twin bug is identical. So we can record the traits of this experiment as ++?, that is, + for A and + for B, but unknown for trait C. Likewise if the outcome is -- for A and -- for B there are still two indistinguishable ways this could have happened as the outcome from a C measurement is forever unknown. Of course the outcome ++? is obviously distinguishable from outcome --?.

As there are two ways to realise the same result, say ++?, Feynman's rule says that first we must add the probability amplitudes for each way, considered separately, then take the square to get the actual probability. Likewise for a --? result. In fact, the same thing must be done in cases where A and B give different results. So a +- or a -+ result for A and B still tells us nothing about what might have happened had we measured C. In total there are four distinguishable outcomes for an A and B measurement: ++, --, +-, -+. For each case there are two, forever unknown, ways to realise the event depending on the two possible outcomes for a C measurement. So the probability for each one of these four outcomes is determined by Feyman's rule. At the end of the calculation the probabilities for each of the four possible outcomes for A and B measurements must still add up to one. However, because Feyman's rule is operating to determine the probability for each of the four events, we are no longer constrained to assume that each event is equally likely and assign a probability of 1/4 to each.

In a moment I will tell you how the probability amplitudes are determined, but here is one result. In one arrangement of analysers, the probability amplitude for ABC to be aligned as +++ is $\frac{1}{4\sqrt{2}}$, however the amplitude for ++- is $-\frac{3}{4\sqrt{2}}$. Notice the minus sign here. Probability amplitudes do not need to be positive numbers as at the end of the day we are going to square them to get a respectable positive probability. Now to get the probability for A and B to give ++ we must add the amplitudes for the two indistinguishable ways this can occur, that is we add $\frac{1}{4\sqrt{2}}$ and $-\frac{3}{4\sqrt{2}}$ to get $-\frac{1}{2\sqrt{2}}$. This is still negative. No problem. We now square it to give $\frac{1}{8}$. A similar calculation shows that the probability for A and B to give -- is also $\frac{1}{8}$. Now we can add these probabilities to get the probability for A and B to give the same result, that is, either ++ or --. These events are distinguishable. They will be written down on the experimenter's notepad quite distinctly, so we can use Bayes' rule with confidence. Thus adding $\frac{1}{8}$ and $\frac{1}{8}$ we get $\frac{1}{4}$. This is the result found in the experiment. The trick is that because probability amplitudes can be negative as well as positive, we can get some cancellation. This is what enables the quantum result to come out as 25 per cent, considerably less than the classical result of 50 per cent.

The probability amplitudes have considerable freedom to take on negative values, but there is still a constraint. The final probabilities that we calculate for each of the

65

$$\frac{1}{6} + \frac{1}{6} + \frac{3}{6} + \frac{3}{6} = \frac{8}{6} = 1$$

mutually exclusive events (++, −−, +−, −+) must add up to one. This means that even as some probability amplitudes cancel to lower the chance of some events, other probability amplitudes tend to reinforce each other to increase the chance of other events. In the case of the last paragraph, we find that while the probabilities for ++ and −− are only $\frac{1}{8}$ each, the probability for +− and −+ are each $\frac{3}{8}$. Some probabilities get smaller as a result of Feynman's rule while others must get bigger.

Where did all those complicated-looking negative fractions with square roots come from? What does determine the probability amplitudes for each way a particular result of an A and B measurement can occur? The answer is that the probability amplitudes only depend on the difference between the two angles that the A and B analysers are set. There is also a constraint acting to ensure that no matter how amplitudes need to be combined, all the probabilities for mutually exclusive events must add up to one. It is not an easy matter to explain how the mathematics actually works. The key is once again to determine the probability amplitudes corresponding to the two distinct, but forever unrealised, outcomes of measurements in the C analyser, and add these two amplitudes according to Feynman's rule.

The bugs exhibit a new kind of genetics, which can be explained only by a new way of dealing with probability. Of course the story is fictional. So far as I know there is no life form that obeys quantum genetics, or at least we haven't found it yet. There is a good technical reason, known as *decoherence*, why quantum genetics is unlikely to ever make an appearance. We will come to the subject of decoherence later. But exactly the same result is obtained for a rather less complex bit of stuff, the photon. Photons are a lot simpler than extraterrestrial organisms and for photons it is very easy to produce twins which obey the laws of the quantum.

Photon twins are produced in great abundance when laser light is passed through a special crystal known as KDP (short for potassium dihydrogen phosphate). We can explain this in terms of the photon picture of light introduced by Einstein in 1905. In this picture, for certain kinds of experiments light is regarded as being made up of a rain of little particles called photons. If we change the colour of the light, the photons have different energy. If we change the intensity we change the number of photons.[3] If green light is directed through the crystal, it comes out diminished and two beams of red light are generated. Inside the crystal some green photons are destroyed and two identical red photons are created in their place, which travel in different directions symmetrically arranged either side of the incoming green beam.

The photons have another property, polarisation, a vestige of the behaviour of light in classical physics. When photons come in great abundance, light behaves like a wave. This wave is, in fact, an oscillating and mutually reinforcing electric and magnetic field. The electric force in the wave can only point in a plane perpendicular to the direction the light wave is moving. In this plane it can have a definite direction, in which case we say it is plane-polarised. To determine the polarisation of light we can pass it through a polarisation analyser. If the polariser is pointing in the same direction as the electric force in the light, all the light is transmitted. If the polariser is at right angles to the electric force, no light gets through. In between are various shades of grey. We can talk of the polarisation of a single photon if we let light of very low intensity through the polariser. Depending on how we rotate the polariser we can get all the photons transmitted, none transmitted or some fraction transmitted. However, if the polariser is aligned at some angle, we cannot be sure which photon will get through: some do, some don't. If there are a lot of photons, this corresponds to the partial transmission of a classical wave at the polariser when the alignment of the electric force in the wave is not the same as that of the polariser. The only time we can be sure if a photon is transmitted is when the polarisation of the photon is aligned to the polariser and the only time we can be sure the

67

photon does not get through is when the polarisation of the photon is perpendicular to the polariser. In this way we can think of a photon as having two mutually exclusive polarisations with respect to a given polariser.

When the photons are produced in pairs inside the crystal it turns out that they always have opposite polarisation, no matter how we align the polarisers. The polarisation of the two photons are totally anticorrelated. This is just the kind of situation we need to realise a quantum entangled state. The photons are produced in pairs. If one goes though a polariser the other does not, with certainty. Every experiment that has been done on these kinds of light beams has produced results entirely consistent with quantum theory and at odds with the classical predication. Entangled states appear to be a fact of life in the quantum world.

Entanglement . . . who cares?

Quantum entanglement describes correlations between the results of local measurements performed on two particles, but the correlations cannot be accounted for in terms of ordinary probabilistic reasoning. By local measurements I mean measurements performed on each particle even if the entangled particles are well separated at the time of the measurements. The results of the local measurements are, however, completely random for a maximally entangled state. Suppose the local measurements have only two mutually exclusive outcomes which we label + or -. Then each result of a local measurement is as random as a coin-toss. To see the correlations we must compare the results of the local measurements on each of the particles. For example, we may discover that if the same local measurement is performed on each particle the results are the same for each particle. Or we may discover that the results are always the opposite for each particle. Of course there is still randomness in these correlations as we only require that the results of the same local measurements are the same. But as there are two results for each local measurement there are two distinct ways for the results to be the same (++ or --). Each of these occurs with the same probability. This level of randomness is essential for defining entanglement, but by itself it is not enough. For example, suppose we mix an urn of red marbles and an urn of blue marbles. You are now told that someone has taken two marbles at random and placed them into a single box subject to the requirement that both marbles must have the same colour. What is the probability that if we draw one marble from the box it will be red? It is not hard to see that this is 50 per cent, even though we know for sure that both marbles in the box have the same colour. We would not call this a case of quantum entanglement.

The key element in defining quantum entanglement is that the probability of the two distinct ways in which the correlation can be realised must be determined by two probability amplitudes which have the same length but which may point in different directions. The fact that they are the same length ensures that the probability for each of the two distinct ways to realise the correlation will occur with the same probability. How then is one ever to see the quantum entanglement? To see the entanglement you must *not* make local measurements of the variable which reveals the correlation, but must rather make local measurements of some other pairs of physical quantities. This is the essential step in realising quantum entanglement and it was first understood by J.S. Bell. Quantum entanglement is thus very like any other puzzling feature which arises from Feynman's rule, such as the interference experiment I discussed in Chapter 1. The big difference is that for a quantum entangled state Feynman's rule is applied to the *joint state* of two or more particles.

How many distinct ways can we quantum entangle two particles with respect to a measurement which has only two mutually exclusive results? There are two ways to specify the correlations which must ensue. We can require the results of local measurements on each particle to be the same, ++ or --, or we can require that the results of such measurement be different, +- or -+. Note that in both cases there are two distinct ways to realise the correlation. The probability for each distinct way to realise the correlation must be the same so the probability amplitudes for each distinct way will have the same length. However, as the probability is obtained by *squaring* the length, there is some further freedom in how we specify the probability amplitudes, that is, they can be either $+\frac{1}{\sqrt{2}}$ or $-\frac{1}{\sqrt{2}}$. For example, suppose we state that the correlations are such that both measurement results are the same, which can be realised in two ways, ++ or --. We thus need two probability amplitudes, one for ++ and one for --, and each amplitude can be $+\frac{1}{\sqrt{2}}$ or $-\frac{1}{\sqrt{2}}$. This would seem to give a total of four different ways to specify this correlation in terms of amplitudes. In fact, there are only two as changing the sign of both amplitudes together will have no overall effect. It is only the sign difference between the two amplitudes that matters. Then we can either have the amplitude for ++ to be the same sign as that for --, or they can have opposite signs. In other words, there are two physically distinct ways to realise the correlation in which both results are the same. A similar result holds for the other kind of correlation in which both results are different (+- or -+). Thus there are four distinct pairs of probability amplitudes that can determine the two kinds of observed correlations. The states corresponding to these four kinds of probability amplitudes are now known as the *Bell states* in honour of the man who first showed us what quantum entanglement really meant. There are *four* distinct ways to realise only *two* kinds of correlations (both results the same or both results different). This doubling of possibilities is due to the freedom that comes from using probability amplitudes to determine a probability. Why should nature come with this extra level of distinction? The answer to that would change our understanding of quantum theory.

So to explain quantum entanglement we are once more back to Feynman's rule and this strange business of prob-

70

ability amplitudes. Why must we resort to this rule? Is there some simple self-evident principle from which Feynman's rule follows? To apply standard probability we simply count things, events, properties, rolls of the dice . . . just things. Yet to account for the correlations in entangled states we must use a different rule for probability. Is this just stretching the notion of probability a little too much? Many think so. We only need to apply Feynman's rule to those events which can be realised in two, or more, indistinguishable ways. Why do we need to do that? If there are hidden variables determining the outcomes to the three possible orientations of magnetisation, why can't we just count them and apply ordinary probability? Well, it seems we cannot. This might suggest that there are no hidden variables at all. Yet how are we to account for the perfect correlations observed in those cases where the same local measurements are made on each particle? In that case we do not get into trouble by postulating hidden variables. Feynman's rule is a way of reasoning with probability for just those cases in which we have no right to invent hidden variables to explain the results. If the results of an experiment are recorded, say an A and B measurement gave ++, even if the results are unknown to me or you, we can apply ordinary probabilistic reasoning. Only when the results are in principle hidden behind the veil of quantum uncertainty can we not apply ordinary probability. What does this tell us about the nature of the quantum world?[4]

CHAPTER THREE

TELEPORTATION FOR GAMBLERS

Mathematics is seductive because it is so precise, but how do you know what you are doing? Where is the reality check? A problem is stated and away you go, working through a series of steps until you are done. Of course there is always more than one way to reach the end, and often experience with previous problems will suggest which path to take. A mathematician knows when she has the problem solved, although it is not enough that she convince herself of the veracity of the solution. She must solve the problem in such a way as to convince her colleagues that the answer is correct, which usually amounts to convincing them that the method of solution is correct. Nonetheless I can't help feeling a bit uncertain about the business. Unlike a problem in physics, a mathematician does not have the template of reality to provide the ultimate justification of a solution.

When I solve some problem in theoretical physics, it is more than just mathematics. The world works in a particular way and the mathematics is constrained to reflect this. If you know what you are doing you can often solve the problem in principle without making too many marks on paper, simply by using your expectations about the way the world works. The details must be produced at some point, actual numbers for the ultimate reality check, an experiment. Every now and then a big surprise turns up. It is confusing for a while, but confusion is good; it suggests something new about the way the world works is lurking just beyond your equations. John Bell turned up a really big surprise when he looked a little deeper into the correlations inherent in the entangled state devised by Einstein, Podolsky and Rosen. In 1989, Anton Zeilinger and Danny Greenberger started thinking about correlations for more than two particles and also turned up a surprise, not by a careful mathematical search, but by thinking about the important physical facts behind EPR and Bell states.

In 1989, Daniel Greenberger of City University in New York City was visiting the Technical University of Vienna, supported by a Fullbright Visiting Fellowship. Early in his visit he got together with Anton Zeilinger to discuss possible projects of mutual interest. They soon settled on the idea of looking for entanglement between more than two particles. One way to do this is by a mathematical search through the Hilbert space of dimensions higher than four (the dimension needed to describe the Bell states). Hilbert space is where quantum entanglement lives, but it is a very big and unfamiliar place, full of strange corners and poorly understood geometry. A brute search would soon lead any investigator into a wilderness of mathematical obscurity. It was not even clear to Zeilinger and Greenberger that an entangled state would exist in these higher dimensional spaces.

Instead of a brute search, Greenberger and Zeilinger began with possible physical mechanisms that might lead to entanglement. A plausible physical mechanism for producing correlations in four particles was soon devised and Danny Greenberger went off to work the math. Unfortunately the results reflected the tangled jungle that is Hilbert space in higher dimensions—too many free parameters and two many correlations to make much sense of. Some surer path would need to be found than simply plodding through the math.

Zeilinger suggested that it might be better to first look for states which exhibit perfect correlations, just like the EPR states of two particles. Such perfect correlations were introduced by Einstein et al., so as to motivate their claim that quantum mechanics would need to be completed by so-called 'elements of reality', or hidden variables. Of course we know now that while such variables may explain the perfect correlations, they cannot explain the cross correlations as discovered by Bell. Zeilinger surmised that it might not be possible for local hidden variables to explain even perfect correlations in states of four particles. Greenberger then returned to the math and soon found that indeed, hidden values do not work for correlated states of four particles. Unlike Bell's discovery, the surprise in the GHZ entangled state would not require a statistical analysis of correlations. If a certain event occurred *at all*, local hidden variable theories would be inconsistent with quantum theory. Zeilinger, while excited by the results, had a feeling that:

> this might not be interesting because we could not see immediately a realistic system where those correlations would show up.[1]

They knew that if their discovery was going to be of any interest to physicists and not remain more than a mathematical footnote to Hilbert space, it would need to have some observable consequences in an experiment. However, together with Michael Horne they presented their results to a conference and subsequently, with Abner Shimony, in *The American Journal of Physics*,[2] a journal devoted to pedagogical aspects of physics. Meanwhile David Mermin from Cornell University, upon hearing of the GHZ result, constructed a simple version in terms of an entangled state of only three particles. His presentation appeared in *The American Journal of Physics* shortly before the paper by Greenberger, Horne, Shimony and Zeilinger. Mermin showed, in a particularly intuitive and convincing manner, that prefect correlations in a three particle system could not be explained by a local hidden variable theory. There would always be a possible result which, if it occurred, would be inconsistent with any local hidden variable theory.

GHZ

When we make measurements on two entangled particles, the entanglement appears as a puzzling cross correlation between the results for local (separated) measurements made on each particle. These correlations run counter to our intuition, but to see it requires a rather detailed analysis of a large number of measurement results. This is very interesting, but it lacks a certain impact. Is there some single-shot, make-or-break experiment we can do for which our everyday counting method for dealing with random events says one thing, but quantum probability says the exact opposite? Is there an experiment for which classical physics says 'yes', but for which quantum physics says 'no' *in a single run*? Yes, there is and the world first learnt this in 1990 when Daniel Greenberger, Michael Horne and Anton Zeilinger presented a scheme, now known as the GHZ scheme. To really force the issue of hidden variables, said GHZ, consider not quantum twins, but rather quantum triplets.

It probably occurred to you while reading Chapter 2 that the problem with trying to survey three different properties on twins is that, in any run, we can only ever measure two different properties, one on each of the pairs. Would it not make more sense to try the same thing on triplets? Surely then we can evade Feynman's rule as we can measure a different property on each one of three highly correlated individuals. It will turn out that not only can we *not* evade Feynman's rule this way, but that the results of quantum mechanics are in even more direct conflict with our classical intuition than in the case of entangled twins.

Let us suppose that in some physical process three particles are always produced. Each individual particle from a particular triplet is then sent through a magnetic orientation measurement (see Figure 3.1). Each device can be set to measure orientation with respect to the direction of the particle's motion or in two other mutually perpendicular directions. For example, we can seek to determine if the

75

Figure 3.1 A source of particle triplets produces three identical particles which then move towards three equidistant magnetic orientation detectors. The detectors can be set to measure the magnetisation along the direction of the particle's motion (the z-direction) or along two other mutually perpendicular directions.

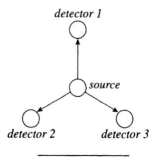

orientation is in the same direction as the motion of the particle or in the opposite direction. Let us call this setting of each detector the A setting. The other two settings, B and C, determine orientation in two mutually perpendicular directions which are at right angles to the direction of travel. In every case when the particles are produced it is found that either all members of a particular triplet have their magnetic orientation parallel to the direction of travel or all members have an orientation that is antiparallel. Thus when all the devices are at the A setting, they either all register a + result, corresponding to orientation along the direction of travel, or a − result, corresponding to orientation antiparallel to the direction of travel. The two mutually exclusive cases occur with equal probability.

We now consider two descriptions compatible with this experiment. The first description is the classical description based on hidden variables, 'genes', for each particle. The second description is the quantum description. It makes use of Feynman's rule so that if a result can occur in two distinct but indistinguishable ways, there is a probability amplitude for each way, considered separately. The first fact we must account for is the perfect correlation that occurs when all detectors are at the A setting. In this case all detectors register a + or all detectors register a –. We represent this by a triplet of results +++ or – – –. It is found that for setting A, these are the only two possibilities and each occurs with equal probability, that is, 50 per cent. There is thus some uncertainty in the outcome despite the perfect correlation present. We can only say that, with respect to the A orientation, all particles have the same orientation, but we cannot know for sure in any given run if this is going to be all + or all –. In the classical description there is no difficulty in accounting for this. We just make sure that in all the possible hidden variables, or genes, for each particle, the A-gene for each member of a triplet is the same. In the quantum description we assume that each of these results are determined by a probability amplitude. There are two amplitudes for determining the outcome of measurements with the A setting. One amplitude (+++) corresponds to the other (– – –), and each amplitude is an arrow of the same length, so that the probability of each outcome is the same. So long as we confine ourselves to A settings, we will never see any contradiction between quantum and classical predictions. However, as the correlations are assumed to be established by two probability amplitudes for A settings, the state of each triplet is entangled. If this entanglement is to make a difference on the outcome of experiments on triplets we need to make measurements when detector settings *other than* the A settings are used. Let us see how it works.

Instead of using the A setting we only use the two settings B and C, which measure magnetic orientation in two mutually perpendicular directions and also in directions perpendicular to that measured in the A setting. The surprising result turns out to occur for experiments in which either:

a   any two detectors have the same setting; or
b   all detectors have the same setting.

To be specific, let us assume that detector one is set to B and the other two detectors are set to C. To see the consequences of entanglement it suffices to consider only a subset of the data in which all the measurements made on particle one (which is set to B) give a + result. There are thus two kinds of experiments we can do. Either all detectors are set to B, which we label as type-1 (T1) experiment, or detector one is set to B and the other two are set to C, which we label a type-2 (T2) experiment. So we can use two triplets of settings:

- T1: B_1 B_2 B_3
- T2: B_1 C_2 C_3

The possible records of experimental results for T1 measurements, given that the results at detector one are +, are given in Table 3.1. If instead we make T2 measurements, and only consider results for which particle one gives a +, the possible results are summarised in Table 3.2.

If the genes in each particle which determine their magnetic orientation are all equally likely, apart from the proviso that the orientation in the A-direction must be the same for all particles, then each of these outcomes should be equally likely, that is each of the four results for a T1 measurement should occur with a probability of 25 per cent. Likewise for a T2 experiment. However, if the statistical states of the three particles are quantum entangled, we find that the first two rows of results in the list of T1 experimental results *never occur* and that the last two rows in the list of results for T2 experiments *never occur*. This is a far more decisive test of quantum entanglement than occurred for an entangled state of only two particles.

Let us consider in more detail just how the assumption of classical hidden variables, which I am calling genes, will determine the outcomes. In this model we assume that each particle has a specific gene that determines how it will respond to one of the three detector settings A, B or C. There must be a total of eight genes for each particle as depicted in Table 3.3.

Table 3.1 T1 magnetic measure

B_1	B_2	B_3
+	+	+
+	−	−
+	+	−
+	−	+

A measurement of magnetic orientation in direction B is made on each particle. Each measurement can have only two results, + or −. For three measurements this means there are eight possible sets of three results. If we only consider those cases in which measurements on particle one give a +, we get the four possible cases shown above.

Table 3.2 T2 magnetic measurement

B_1	C_2	C_3
+	+	+
+	−	−
+	+	−
+	−	+

A measurement of magnetic orientation in direction B is made on particle one and in direction C on the other two particles. If we only consider those cases in which measurements on particle one give a +, we get the four possible cases shown above.

Note that the first four rows in Table 3.3 correspond to an A-detector registering a + while the last four rows correspond to an A-detector registering a −. If quantum entanglement is present, we can expect to get into trouble when combining these cases if we never measure the A orientation.

Now when triplets are created each particle has a gene which has the same A variable, though the other two positions are completely random. This ensures the perfect correlation observed for A measurements is faithfully

Table 3.3 Genetic combinations for three variables where the A is the same for all three particles

GENE	A	B	C
G1	+	+	+
G2	+	+	−
G3	+	−	+
G4	+	−	−
G5	−	+	+
G6	−	+	−
G7	−	−	+
G8	−	−	−

A summary of the eight possible genes, or value triplets, which determine the outcome for each of the three possible measurements A, B or C we can make on each particle. The only restriction we impose is that, for a given triplet of particles, the A-gene for each particle is the same. However, we will never measure the A-gene. We will only measure combinations of B and C on each particle in triplet.

explained. To calculate probabilities, however, we need to start counting a lot of genes. There are eight genes for each of three particles, so that would give $8 \times 8 \times 8 = 512$ different possibilities. We can get this down to something a bit more manageable. First, the A-gene for each particle in a triplet must be the same. Second, we are going to restrict ourselves to explaining just those data for which the result at detector one was a +. For example, suppose we ask for the total number of gene combinations in which a T2 experiment (that is, $B_1C_2C_3$) could give +++. There are eight gene combinations which would give this result in which the A-gene of each particle is a + and there are eight combinations which give this result in which the A-gene of each particle is a −. These are given in Table 3.4.

All the gene combinations in Table 3.4 are equally likely so there are sixteen ways for a T2 experiment to give +++.

Table 3.4 Eight genetic combinations for T2 measurements where the result is +++ and A is constant

A = +	A = −
$G_1G_1G_1$	$G_5G_5G_5$
$G_1G_1G_3$	$G_5G_7G_5$
$G_1G_3G_1$	$G_5G_5G_7$
$G_1G_3G_3$	$G_5G_7G_7$
$G_2G_1G_1$	$G_6G_5G_5$
$G_2G_3G_1$	$G_6G_5G_7$
$G_2G_1G_3$	$G_6G_7G_5$
$G_2G_3G_3$	$G_6G_7G_7$

A list of the combinations of genes for each triplet which give a result (+++) for a T2 measurement. In every case the A-gene is the same for all three particles in a triplet, but could be + or −. The genes are arranged according to whether or not the A-gene is + or − for each triplet.

In a similar way we find there are sixteen ways for each of the other three results for T2 measurements to occur, and likewise there are sixteen ways for each of the four possible results for a T1 measurement. So each of the four outcomes for a T1 measurement must be equally likely as must the four outcomes of a T2 measurement.

To put this more precisely, we see that there are four possible outcomes for experiments of T1 (Table 3.1). Each of these can occur in sixteen ways, so that there are a total of $4 \times 16 = 64$ ways to get a possible result for a T1 experiment. Now there are eight ways to get, say, the result +++ given that the A-gene is a +. Thus the probability for this result is $\frac{8}{64} = \frac{1}{8}$. Likewise the probability for a T1 measurement to yield the result +++, given that the A-gene is a −, must be the same, that is, $\frac{1}{8}$. So the probability to get the result +++ for the T1 measurement is just $\frac{1}{8} + \frac{1}{8} = \frac{1}{4}$, as the T1 measurement does not distinguish the + A-gene from the − A-gene. All we know is that whatever the A-gene is, it is the same for all three particles in a

81

given run. So we see that the probability of 25 per cent for +++ comes by a direct application of Bayes' rule when we add the two indistinguishable ways a T1 result can give a result +++. In the case where the A-gene state is quantum entangled, however, we expect this is going to be a problem.

Now to the quantum explanation. We will again only seek to explain those results for which measurements on particle one gave a +. If the states of the A-gene are entangled we never get the results +++ or +-- when we do a T1 experiment, in contrast to a classical statistical description which would imply that each of these two cases occur 25 per cent of the time. In the case that the states of all three particles are quantum entangled in the state of the A-gene, we assign a probability amplitude for all particles giving +++ when all detectors are set to A and a different probability amplitude for the result ---. These are the only amplitudes we need, as our stated correlation requires that all A measurements give the same result on each particle in a given triplet, but we do not specify if each result is a + or a -. Recall that the probability amplitude is a little arrow in two dimensions which can point in any direction. The length of the arrow squared gives the probability. As we only get +++ or --- we conclude that the two probability amplitudes for these events must have the same length to ensure that the probability for each of these two mutually exclusive outcomes are equal.

Now suppose that, instead of doing an A measurement on every particle, we do a T1 experiment on an entangled state, which corresponds to the settings $B_1B_2B_3$. In this case we have no idea what the result of an A measurement would have been, other than that they would have all been + or all been - corresponding to the way in which we have entangled the A-genes. For example, suppose we seek the probability to get a +++ result for a T1 setting. There are two indistinguishable ways to get this result corresponding to the two possible values for the A-gene. Feynman's rule now tells us to first find the amplitudes for each way considered separately, and add them before squaring to find the length. Now it turns out that the two probability amplitudes to get +++ corresponding to whether or not the A-gene is + or –, have the same length *but point in opposite directions*. They will thus cancel completely, so the probability for this event is precisely zero. We never get a +++ result for a T1 measurement. Incidentally the squared length of each probability amplitude is in fact $\frac{1}{8}$, which corresponds to the classical probability for each conditional event. However, in the quantum case we must first add the amplitudes before squaring to find the length and in this case there is a complete cancellation. Quantum entanglement of three particles has resulted in a complete contradiction between quantum and classical reasoning based on hidden variables or 'genes'.

There are a number of proposals for such an experiment. One of the more interesting proposals was made by Seth Lloyd of MIT. Lloyd's scheme is based on the well-developed technique of nuclear magnetic resonance (NMR), the fundamental physical phenomenon underlying the widespread medical instrumentation of MRI (or magnetic resonance imaging). It is particularly intriguing as, in principle, it would enable quantum entanglement to be done in a cup of coffee!

Lloyd proposes to use a particular quantum property of nuclei called spin. Spin has the required binary property that it is either up or down and produces as a result a similar binary behaviour in magnetic orientation. These spins reside in a molecule and while it is not possible to perform macroscopic measurements on individual spins, it may be possible to use another spin in the same molecule as a kind of apparatus. In practice there are vast numbers of molecules, all in slightly different states due to the disorder present at any finite temperature. The trick is to try and get enough of them to cooperate in a GHZ experiment so that a signal can be extracted from a large number of tiny GHZ experiments in parallel. In June 1997, this experiment was successfully carried out by Raymond Laflamme, Manny Knill and co-workers at Los Alamos

National Laboratory. They used the proton and carbon spins of the molecule trichloroethylene. ⌋

Teleportation

Correlations between three entangled particles provide a very clear contrast between quantum entanglement and a classical explanation of correlations based on unknown hidden variables. Can we put this knowledge to some practical use? Perhaps there is a scheme for more efficient communication similar to the ideas in Chapter 2. In March 1993, a paper appeared in the *Physical Review Letters* with the unlikely title of 'Teleporting an unknown quantum state via dual classical and Einstein–Podolsky–Rosen channels'.[3] As you will shortly see this is a very descriptive title, but that word 'teleporting' sounds a bit unusual. It conjures up images of Captain Kirk and Spock. Well you won't be too surprised to learn it is not quite like that. Nonetheless what this paper did show was just as surprising: that by local measurements and classical communication a particle in a state described by Feynman's rule could be measured at one location and another particle at another location transformed into precisely the same state upon receipt of the classical information. Like all quantum proposals this one made use of the strange nature of quantum randomness, so before we get too carried away it might help to look at a classical 'teleportation' scheme . . . for gamblers.

Suppose we have a pair of ordinary (not entangled) identical twins in some distant large city. We will use only a single property, with two mutually exclusive values, to characterise the individuals. To be specific we will consider only if an individual is male or female. For identical twins there is clearly a strong correlation with respect to this property: either both twins are male or both twins are female.

In the population at large, however, it is found that the distribution of males and females is not equal. For some reason, in this city, there are many more females than males. Perhaps all the males stayed back on the farm while the females moved to town. Whatever the explanation we will assume that the probability that a person selected at random is a male is given by P_M while the probability that a person selected at random is female is P_F and that P_M is not equal to P_F.

All the identical twins in this city are enrolled in a program to test the 'twin paradox' of special relativity. One individual from a pair of twins is put in a rocket and launched towards a distant galaxy. The other twin stays at home and may eventually get married. In the interests of equity an equal number of males and females are launched into space. Thus on average the probability that a male is in space is 50 per cent, equal to the probability that a female is in space. There are a large number of rockets heading towards a distant galaxy, each carrying a single person, but overall there are equally as many males in space as females. Of course if a male is launched, his sibling back home is also a male as we are assuming identical twins.

Now suppose we ask the following question. What is the probability that a twin that stayed at home and got married has a brother in space? You might have guessed that this is just 50 per cent, but you would be wrong. In fact the answer is P_F, that is, the same as the probability to find a female in the general population. Somehow the probability to find a female in the general population has been 'impressed' upon the population of siblings lost in space.

Here is the reasoning. Let us assume, not unreasonably, that a twin that stays at home marries a person of the opposite sex. We are only told that the twins that stay at home may get married, so this is the most we can say about their marriage partners. Let us label the twin in space as T_S and the twin back on earth as T_E while the marriage partner of the latter we label S (for spouse). We can now enumerate all the various possibilities in a form like Table 3.5.

Notice that entries such as +–+ do not appear as the twins must be identical and thus will have the same sex. This strong correlation is crucial. The probabilities for the

Table 3.5 Marriage combinations for earth twins and their probability

S	T_E	T_S	Probability
+	+	+	0
−	+	+	P_F
+	−	−	P_M
−	−	−	0

This table shows the number of ways in which a married twin can occur, together with the probability for those combinations. A − signifies a female, while a + signifies a male. In this table we exclude all cases where a twin does not get married at all. Very rarely a twin marries a person of the same sex, but this occurs with a probability approaching zero.

various entries are easy to calculate. There are an equal number of male twins and female twins as we have assumed that the space program always launches an equal number of males and females. However, there are an unequal number of males and females in the general population. If a male twin gets married he must choose a female, which occur in the general population with a frequency near P_F. It is now obvious that if a twin gets married the probability that his or her sibling is male is P_F and not 50 per cent. If all we know is that a twin gets married we cannot say for sure what the sex of the sibling is. Of course there may be many twins who do not get married at all. It will be impossible to verify these statistics until the twin launched into space gets back to earth. This, however, would require that the city authorities keep very good records as it will be a very long time (Earth time, that is) before a spacebound twin gets back. Remember this was a scheme to test the twin paradox.

Another way to check the result is to transmit the information on the marital status of earth twins to the distant galaxy towards which the spacebound twin is heading. The reception committee on the distant galaxy will receive this information. In all those cases for which a message is received a fraction of roughly P_M of arriving twins will be female and in a fraction of P_F the arriving twin will be male. However, overall they will find an equal number of males and females. Only if they exclude those cases for which no message is received do they get a different result.

I am sure you have anticipated what is coming next. Instead of using twins, with their rather ordinary kind of correlation, what happens if we use quantum entangled pairs of particles? This is precisely the scheme proposed in the 'teleportation' paper in *Physical Review Letters*. It works much the same as the twin experiment discussed above, but with one big difference: what gets impressed upon the distant particles is not a probability but a probability amplitude.

Now the scene changes to introduce the dynamic duo of quantum magic, Bob and Alice. Bob and Alice work in widely separated laboratories, and to communicate send qubit particles to each other from time to time. Suppose Bob prepares two particles in a quantum entangled state of correlated magnetic orientation. The particles he uses have only two possible magnetic orientations, up or down, which we will label + or −. The states are perfectly correlated so that whenever one particle is found to be + the other is found to be + and when one particle is found to be − the other is found to be −. Each pair of results ++ or −− occurs with equal probability and, as we found in Chapter 1, there are two physically distinct ways in which Bob can prepare this entangled state. He can either have a state in which the two probability amplitudes for ++ and −− have the same sign or they can have different signs. To be definite I will take the case in which they have the same sign. In Chapter 2 we saw that by measuring magnetic orientation in different directions on each particle of a pair, unusual correlations would appear as the hallmark of quantum entanglement. Now Bob sends one of the EPR particles to his colleague Alice in a distant laboratory and keeps the other particle at home in his lab (see Figure 3.2).

Figure 3.2 Bob sends a pure quantum state to Alice by sending an ebit over a quantum channel and a bit over a classical channel. An ebit is a single bit of information entangled with another single bit.

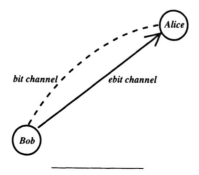

If Alice performs magnetic orientation measurements on her received EPR particle, she will obtain + and − completely at random, regardless of which direction of orientation she chooses for her measurement. All EPR pairs behave this way under local measurements. Of course when we compare the results of Bob's local measurements with Alice's local measurements we will discover perfect correlations for just those cases in which Alice and Bob make measurements of the same orientation. But on their own, the local measurements at either end are simply a coin-toss, just as all twins arriving at the distant galaxy are equally likely to be male as female.

Bob now also prepares another particle, call it the control particle, with a magnetic orientation aligned in direction A with certainty. However, if Bob chooses to measure in another direction, say B, he cannot be sure of the result. It could give either up + or down −. Suppose it has a probability P_+ of giving a result + when it is measured in the B-direction and a probability of P_- to give a − result when measured in that direction. We do not assume that P_+ is equal to P_-. Each of these results is determined by a probability amplitude. These amplitudes must have different lengths (as P_+ is not equal to P_-), but they may also point in different directions. Our objective now is not simply to impress the probabilities for the control particle upon the distant target particle, but rather to impress the *probability amplitudes* of the control particle upon the distant particle. Here is how to do it.

Bob must measure a joint property of the control particle and the EPR particle he kept in his lab. In the case of twins, the joint property we used was 'marriage status' of a pair of particles. This selected out only those pairs which had the opposite sex, but did not tell us what the sex of the earthbound twin in the marriage actually was. Bob will try a similar ambiguous joint measurement. Bob decides to make a measurement which will say YES if the two particles in the lab are measured to have the same magnetic orientations in the B-direction and NO if they have the opposite magnetic orientation. In each case there are two ways to realise each result. A YES result could occur if both particles were ++ for a B-direction measurement or −− for a B-direction measurement. A NO result could occur if the particles were either +− or −+ with respect to a measurement in the B-direction. After the measurement we know for sure if the two particles are YES states or if the two particles are NO states, but we still do not know for sure the magnetic orientation in the B-direction for each particle. If the result is NO, we can only say that the two particles have the opposite orientation in the B-direction. As for the previous example with twins, if we only know that a twin got married we do not know the sex of the twin or the spouse, we only know that they have the opposite sex. In Bob's experiment, the underlying state of the particles after the measurement is still a quantum superposition of the two possibilities. It is a qubit composed of the two ways in which each result YES or NO can be realised.

To be specific let us suppose the result of Bob's measurement was NO. Bob now knows that his control particle has the opposite magnetic orientation to the EPR particle that remained in his lab. This pair of particles is now itself in an entangled state. It is a state with perfect correlations (anticorrelations to be exact) which can be realised in two indistinguishable ways. It is a state described by a probability amplitude for +− and a probability amplitude for −+. Bob keeps records of only those pairs of particles that give the NO result.

In the remote laboratory Bob's colleague, Alice, measures the orientation of arriving particles with respect to the B-direction. She will find that there will be very nearly an equal number of plus results as minus results. It looks just like a coin-toss. However, for some particle pairs, Bob transmits the NO result of his ambiguous measurement to Alice. Now suppose Alice only makes a measurement on those particles corresponding to a NO result from Bob. She will find that in roughly P_+ of cases they will give a + result and in P_- cases they will give the result −, which were the same odds that Bob obtained for B measurements on his carefully prepared control particle. This follows with exactly the same reasoning we used in the case of space-travelling twins.

However, there is now a big difference. To see it, remember that Bob actually prepared his control particles to have a definite orientation in some other direction A. What result would Alice obtain if, upon receipt of the NO message from Bob, she measured the orientation of the received particle in the A-direction? The surprising fact is that for all those particles, she will get exactly the same result. A coin-toss has become a certainty. In other words, the state of the subset of particles indexed by the NO message from Bob is precisely the same as the state of the control particle carefully prepared by Bob in his laboratory. By sending one entangled particle and one classical bit of information, Bob has teleported his control particle to Alice.

Teleportation indicates that entangled particles could be a valuable communication resource when we wish to send qubits rather than ordinary bits of information. Charles Bennett of IBM has suggested we use a different name, 'ebit', for information contained in an entangled state of two particles with binary properties such as magnetic orientation. A new name is a good idea, as we know that there is something distinctly new about the information contained in an entangled state. The EPR states we have been using for teleportation can be said to carry a single ebit. The teleportation thus involves one ebit and one ordinary bit, which represents the information sent over the classical communication channel from Bob to Alice. The discovery of ebits and qubits is leading to a reappraisal of old ideas on communication channels. Will there come a day when Telstra or AT&T offer you the choice of an ebit channel to send your credit card number over the Internet?

CHAPTER FOUR

REALITY, BY NINTENDO™

The United States Department of Energy is betting on the Church–Turing principle. I hope they get it right. Our future security may depend on it. On 16 December 1996, Intel Corporation, in collaboration with the Department of Energy (DOE), announced that the recently constructed ultra-computer was the first computer to reach one trillion operations per second, one teraflop in technical terms. That is a lot of calculations. If every man, woman and child in the United States was given a hand calculator and asked to work non-stop for 125 years they would accomplish what the ultra-computer calculated in merely one second. To achieve this cost $55 million. To what end you might ask?

Were it not for the feasibility of an ultra-computer, President Clinton's signing of the Comprehensive Test Ban Treaty on 25 September 1996 would not have been possible. The ultra-computer is housed in a large room at the Sandia National Laboratory in New Mexico. It is certainly not a desktop machine, though it is built out of the same Pentium Pro processors that power your personal computer. The full system will consist of 76 large cabinets with 9072 Pentium Pro processors and nearly six billion bytes of memory. The machine is the most tangible product of the Department of Energy's Accelerated Strategic Computing Initiative (ASCI). Ultimately the computer will reach a peak performance of 1.8 teraflops, or 1.8 trillion floating point operations per second. Why would anyone want such a computer? Is this just another manifestation of macho geekdom? Well, yes, I suppose, but there is more. The ASCI is a ten-year program to move nuclear weapons' design and maintenance from being test-based to simulation-based. In other words, cyberspace has just gone nuclear, but the bangs are virtual. The consequences are very significant. The Comprehensive Test Ban Treaty is one such consequence.

Would you buy a car right off a specification sheet, a car that had never been built and never been tested, but one that had driven a million miles in cyberspace in perfect safety? Suppose I said this was the safest car in cyberspace. In over one million front-end collisions not a single (cyber-) person had suffered an injury. Would you buy this car? Perhaps you would if the car had been simulated on the ultra-computer. If an ultra-computer can perfectly simulate nuclear tests, it can do a lot more besides. After all, that is the point of a computer. It can be programmed to solve a multitude of problems. I am sure airline manufacturers would love to get their hands on an ultra-computer. How would you feel as the test pilot on the maiden flight of a plane that had only ever flown in simulation? Just how well placed is our faith in computer simulation? Is it possible to perfectly simulate reality on a computer? I am convinced the answer is yes. This is a statement about the way the world actually is. It is a very important principle and deserves a name. David Deutsch, a theoretical physicist at Oxford, has called it the Church–Turing principle. If the Church–Turing principle is right, it is telling us something quite profound about the nature of reality. If the Church–Turing principle is wrong, then a bomb that explodes in cyberspace may be a complete dud in reality. Apparently the United States Department of Defence has no doubt at all about the Church–Turing principle. How can they be so sure?

If you have ever done any computer programming yourself you might be reluctant to endorse the DOE's confidence. As we all know, programs come with 'bugs'.

A software bug, however, is not a fact of reality, despite your frustration every time your new program crashes. The laws of physics do not require that every program contain at least one bug. New techniques in software engineering and verification have enormously improved the quality control on large software projects. We need to be quite clear about this, as when I ask if the perfect simulation of reality is possible, I mean possible in principle. We are going to assume that a computer capable of a perfect simulation of reality can be programmed 'bug-free'; a tall order to be sure, but one that, in principle, we can aspire to. Perfect programming is possible in principle. It is not ruled out by the laws of physics. (Murphy's law is not a physical law.)

But is perfect simulation possible in principle? Might it not be the case that reality cannot be perfectly simulated by a machine we call a computer? Perhaps the world is just too complex to be ever captured fully by the calculations run in a computer, no matter how good it is. To address this issue we first need to ask if a computer can perform all possible calculations. Is all of mathematics reducible, after all, to some suitable code running on the ultra-computer? Perhaps the mathematics that computers can do is not good enough to simulate physical reality. There are two questions to be answered. First, does mathematics itself place constraints on what can be computed; and second, does physics place constraints on the operation of the particular physical devices we call computers? Let us take the first question.

Turing and computability

The issue of what a computation actually is was addressed by the English mathematician Allen Turing in 1936. A similar question was posed by two other mathematicians at about the same time, an example of the curious synchronicity in the tide of human intellect. Alonzo Church, an American mathematician, put forward a mathematical system that addressed pretty much the same issue, as did Emil Post, a Polish–American logician. Remarkably all these approaches were soon shown to be equivalent. Of course there were no computers in 1936, so Turing put the question a little differently. He asked, what are the possible mathematical processes that can be carried out in computing a number? Such a process we call an algorithm. In 1936, the only physical object capable of running an algorithm of any significance was the brain of a mathematician. In a sense Turing was trying to formalise how mathematicians think. Of course this leaves open the question as to what extent a human brain is equivalent to a computer.

Turing's definition of computability is entirely mathematical, it tells us nothing about the world as it is. However, to describe his approach he used a very suggestive image of a machine, now called a Turing machine. In 1936, mathematics was done by human brains reading and writing marks on paper. However, it is as well to keep in mind that Turing was not describing a physical computer when he proposed his definition of computability. A Turing machine is not a physical machine, although we can and do assemble physical systems that behave just like a Turing machine. Nonetheless it is probably true that behind his approach to computability, Turing had an intuition about a classical (that is, non-quantum) clockwork machine.

There are many ways to make marks on paper. We could use the English alphabet, or Chinese characters, and write everything out in standard words. This is not a very efficient way to do mathematics. The decimal number system, supplemented with various symbols, is the way it is usually done by mathematicians. Of course we could use any symbols we like as long as we all agreed on how to decipher the code. To describe how a Turing machine marks paper however, it is much more efficient to use just two symbols, 0 and 1, the binary system. If we rule our paper up in an appropriate way, say as a set of rows and columns like a spreadsheet, we can actually get away with one symbol, by taking for granted that a blank cell codes for 0. Let us suppose we have a long paper tape ruled up into little cells. Each cell either contains a 1 or nothing, in which case it is said to contain a 0. We are going to do all our calculations by reading and writing and erasing marks in cells on a very long, possibly infinite, paper tape (see Figure 4.1).

In addition to the tape, we have a 'head' which can be in a finite number of different states, each labelled by a single number (see Figure 4.2). The head could be a real mathematician's head, or it could be some kind of computer. What physical manifestation it takes is not important. What matters are the *rules* that the head uses to read and write to the tape. That said, of course, the words we are using are highly suggestive, and we will need to ask if we have introduced hidden assumptions that unnecessarily restrict our concept of computation. A particular concern is the notion of the head 'state'. For now, we just assume that there are distinctly different states of the head, distinguished by a single number. The intuition that lurks behind this, however, is some kind of classical physical device with identifiable and measurable properties that enable the states to be distinguished. As quantum mechanics shows, this intuition is wrong.

The rules to read, write or erase marks on the tape need to depend on at least three things: where the head is, what mark is being read, and what state the head is in.

Figure 4.1 The paper tape of a Turing machine contains a number of 'cells' which either contain a mark, indicated by a 1, or are blank, indicated by 0.

The head is only allowed to address a single cell on the tape at any one time, so we can dispense with a locater. It just does whatever it needs to do to the cell on the tape beneath it. The head can read the cell and do one of three things to it: nothing, erase it (that is, change a 1 to a 0) or write a 1 (if it is blank). Having done one of these three things, what does the head do next? It may move one cell to the left, or one cell to the right, or it may not move at all, that is, it stops. What the head does to the tape, and how it moves, is determined by what state the head is in when it reaches a particular cell. We also must allow the state of the head to change, after it acts on a particular cell. This captures the idea that what a mathematician

Figure 4.2 Representation of a Turing machine. A paper tape containing the data and the program is being read and modified by the head, moving on a pair of rails. The number labelling the head state is displayed in a window on the front of the head. In the figure this is the binary number 1010. Only a small segment of the tape is shown. In principle the tape may be extended indefinitely.

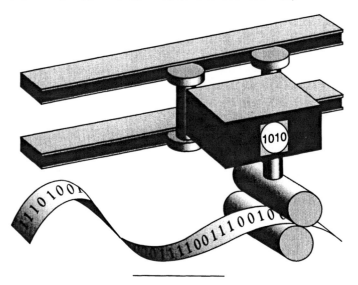

decides to do next depends on what she has just done, and introduces a necessary conditioning of the machine's action on its past history of action, conditioning essential to the notion of computation. The head will continue to read, write, erase marks and move backwards and forwards along the tape until it reaches a situation where its internal state, the square it is currently reading, and the symbol on the square remain the same from step to step. In that case the machine is said to 'halt'.

Here is an example. Suppose we want to simply multiply a number by two. Let us take the number to be four. The first thing we need to do is to code the input as a binary number so that it can be represented on the tape as a set of marks and blanks. In the binary number system we can count from zero to one in the usual way. However, when we get to two, we need to do something else, as we only have the symbols 0 and 1 to work with. We all know how to do this with the decimal number system. We have ten symbols, 0, 1, 2, 3, 4, 5, 6, 7, 8 and 9. Now to represent the number ten, we start over again one notch to the left. We use 1 followed by 0 to give 10. Another way to put this is to say we put a 1 in the tens column and a 0 in the ones column. Then we just increment the first digit again. The same trick works for binary numbers. So to count up to three we have 0, 1, 10, 11. To represent a two we put a 1 in the twos column and a 0 in the ones column. To go to the next stage we add another notch or column, the fours column. So the number five is one in the fours column and a one in the ones column to give 101. Each column is a power of two, just as in the decimal system each column is a power of ten. Note that all even numbers must have a zero in the ones column.

Now when we multiply four by two we get eight. In the binary system we write 4 as 100 and 8 as 1000. It is easy to see that all we need to do is insert a zero into the first position and move everything else one notch to the left. This will work with any number, as any number multiplied by two is even and thus must have a zero in the ones column. We need to come up with a set of rules so that no matter what binary number is coded on the tape, it will be transformed in just this way, that is, a zero inserted at the right. The picture of computation that is emerging is one in which one binary string, the input, is transformed into another binary string, the output. We need some convention as to where the input on the tape will be. Let us suppose that all our input binary strings are fed in from the left of the head and when the calculation is complete the output is the binary string that lies on the left of the head. The head then starts just to the left of the binary string that encodes the input and finishes just to the right of the string that encodes the output.

Now we can give the set of rules that define the Turing machine for multiplication by two.[1] To do this we need a Turing machine with four distinct head states. We can label these 0, 1, 2, 3. In keeping with our binary coding, however, we will also convert the head state number to binary form. We will do this by 0→0, 1→1, 2→10, 3→11. For each head state there are two possible symbols on a tape cell that the head can read at any one step. The Turing machine for multiplication by two is now defined by a set of conditional transformation rules that tell the head what to do, depending on what symbol it is reading and what state it is in. They are:

00 → 00R, 01 → 10R, 10 → 01R, 11 → 100R, 100 → 111R, 110 → 01STOP

The first symbol on each side of the arrow is the label for each head state. The second symbol, in bold type, on the left of the arrow is the symbol on the particular cell that the head is reading. On the right hand side of the arrow, the first symbol tells us what state the head must change to, the second symbol (in bold type) tells us what symbol the head leaves on the tape, and the final symbol tells us how to move the head, R for one step to the right, L for one step to the left (not used) and STOP. It is now a simple matter to write down a sequence of rules from the list above, to perform multiplication by two. I invite you to try.

Of course more complicated calculations, such as division, will require more head states and a much larger list of transformation rules. You should try some examples for yourself, perhaps adding two numbers. The list of transformation rules will grow quite rapidly. Despite this, after a while you should convince yourself that really any calculation at all can be done this way.

A computer, however, is not purpose built to solve just one problem. We need to be able to specify what calculation we want to do, together with the input. In other words we need a program and the data. However, a program itself can simply be a set of instructions coded as a binary string. It is not difficult to design Turing machines that read a binary string which specify which transformations to apply, and then act according to those transformations on the input data. It is even possible to arrange things so that which rules get applied depend on the results of intermediate calculations. It soon becomes clear that any mathematical operation at all can be performed by a suitable Turing machine. What is a suitable Turing machine? It must be one with a sufficient number of head states and transformation rules to be able to do any computation at all for a given input of program and data. This is the universal Turing machine. Such a machine is, of course, not a real machine, but simply a set of transformation rules. Those rules themselves must be sufficient to be able to reproduce any other Turing machine, such as our example of multiplication by two. The universal Turing machine can simulate any other Turing machine. All we require is to input the binary code for the particular Turing machine we wish to simulate along with the program and data on which it is to act. The transformation rules of the universal Turing machine can also be coded in binary form. The actual form you get will depend on a number of details of how this coding is done. The result is typically a rather long binary string.[2] Now we have a way to formalise what

100

we mean by an algorithm or computation. This is the Church–Turing thesis:

A computable function is one that is computable by a universal Turing machine.

We have considerable confidence that this does indeed define what we mean by computable. The confidence derives in part from other completely different approaches, such as that of Church and Post, that led to equivalent definitions. In addition we also have programmable computers which, given enough memory resources, can act exactly like a universal Turing machine. The big question is, are there mathematical functions which do not fit our definition? Are there mathematical functions that are not computable?

The answer is yes. To see that we must ask: How could a Turing machine fail to compute a particular mathematical function? The answer is really quite simple. If the universal Turing machine never stops, it cannot be said to compute anything. So the question becomes: Are there problems for which a universal Turing machine would run forever in trying to solve? This has become known as the 'halting problem'. Here is an example, the domino snake problem.

Suppose we have an inexhaustible supply of dominos. The dominos are a set of tiles carrying different symbols, usually an array of dots, on each end. We are going to place these dominos on a lattice of squares. A chess board is a lattice of squares. Let us assume that our dominos are each just the right size to fit snugly over one or two squares on the chess board. We are only allowed to place dominos together on the chess board if the touching edges carry the same symbols. To begin, place two randomly chosen dominos, A and B, anywhere on the lattice.

Can we connect these two dominos by a domino snake, following rigorously the rules for placing the dominos? If the chess board is finite, the problem obviously has a definite answer, yes or no. If the chess board is infinite

101

and the dominos can go anywhere, the problem is also decidable, and the time taken to solve it does not get too big even if A and B are quite far apart. Now the interesting result is this. If you take an infinite chess board, but draw a line on it which the domino snake is never allowed to cross, the question as to whether we can join the two dominos is undecidable. (Of course we must place A and B on the same side of our impenetrable line.)

There any many examples of problems which are undecidable. How do you figure out if a given problem is undecidable? In any particular case, such as the snake problem, the answer is either yes or no, and this can be determined by sufficient ingenuity. Suppose we code the problem as some long binary string. The input is a program and input data string, the output is the answer, yes or no. Let us code yes as a 1 and no as a 0. Now we have a mathematical function which takes the binary number representing the input and produces a single binary number as the output. Is this function itself computable? Suppose we code the problem up as a long string of binary numbers and hand it to our favourite Turing machine. If the Turing machine comes to a stop, it will have produced the answer (yes or no) to our problem somewhere on the tape. However, suppose the Turing machine never stops? In that case we can say the problem is undecidable. But how do we know in advance if this particular Turing machine applied to this particular problem is going to halt or run forever? Is this problem decidable by a universal Turing machine? Turing proved that the answer is no. There is no universal Turing machine that will determine if a given Turing machine, working a particular problem, is going to stop. In some cases a universal Turing machine may run forever and produce an apparently endless string of ones and zeros. To any outside observer that string would be indistinguishable from an endless coin-toss.

There is no general algorithm which will determine if a Turing machine working on an arbitrary input is going to finish and turn itself off, or run forever. This is the

102

content of an important theorem know as <u>Turing's theorem.</u>
It would make it much easier to manage my time in front
of the computer if Turing's theorem were not true. It would
be nice to have a little application running on my machine
which took every program I wrote and printed out how
long it would take to run, before it actually ran it. I might
then decide that I have enough time to go out for a cup
of coffee. I am sure my graduate students would like to
know if the problem I have set them for their thesis is
going to finish or run forever. But there is no way to give
a guarantee that a problem will finish in general. The best
you can do is sit down and start to solve it, or program
your computer to do it for you, and go out for coffee
anyway.

Reality 2001: the world by Microsoft™

We see that mathematics does indeed place constraints on
what we mean by computable. Now we can return to our
initial question: Is reality itself computable? Is a perfect
simulator of reality possible in principle, or is the physical
world beyond what we can ever simulate on a universal
Turing machine? Can we expect Microsoft to ever produce
a super program, *Reality 2001*, which literally gives you the
world? According to David Deutsch, the answer is yes, but
Bill Gates will need to brush up on his quantum theory to
do it.

David Deutsch, a theoretical physicist at Oxford University, has stated the Church–Turing principle as follows:
'Every finitely realisable physical system can be perfectly
simulated by a universal model computing machine operating by finite means.'[3] There are a few words in there,
such as 'finitely realisable' and 'universal', that will take
some explaining. However, the most important word is
missing because Deutsch takes it for granted. That word
is 'quantum', for a universal model computer *must* be a
quantum computer. As we shall see a quantum computer
is nothing like a classical computer on which all of computer

103

science has thus far been based. The Church–Turing principle is only true if we can build a quantum computer. If we can build a quantum computer then the ultimate virtual reality generator can in principle be built.[4]

The claim is that *Reality 2001* perfectly simulates all of reality on the hyper-computer. How can we test this claim? Before we can answer that question we need to ask, how do we test the real thing? How do we test reality itself against competing explanations? That of course is the business of physics. The physical world is a complex place, displaying regularity amid a background of chance and randomness. We can learn a lot about the world just by looking at it, but appearances are deceiving. In the sixteenth and seventeenth centuries a new way of comprehending reality was devised based on experiment and mathematics. The idea that anything at all can be learnt about nature by artificially constraining it took a long time to take hold. Aristotelian natural philosophy would have found the idea quite alien. It is now taken for granted that the only reliable path to knowledge of the physical world is through the experimental method.

All our theories about reality are based on a finite set of measurement results. There is a widely held, and entirely misguided, belief that the purpose of theory is simply to provide some order to an otherwise random assortment of experimental data. You often hear this view expounded even by physicists, who should know better. In this view, all that matters is that we get the right numbers out of a physical theory. The right numbers mean those that agree with experimental results. It is of course true that theories should predict the right results for measurements, but they are much more than that. Physical theories provide *explanations* that capture our *understanding* of reality. Physics is much more than just mathematics. A physical theory introduces concepts and provides relationships between them. Usually these relationships are mathematical. Another idea with some currency in contemporary discourse is that physical theories are simply 'stories' told by physicists to each other.

This view is often expounded by Literary Theorists, a species of University Academic recently evolved through the selective pressure of Grant Applications. It is easy to demonstrate that this is not a correct view and I shall return to it later.

Ultimately a physical theory must produce numbers which are confirmed by measurement results or at least consistent with them. Both measurements and physical theories produce numbers, but in entirely different ways. Physical theories produce numbers through mathematics. Measurements produce numbers through experiments on the real world. It is a very significant fact, not unconnected with the Church–Turing principle, that the numbers produced by measurements can also be obtained by mathematical theories of reality. It is indeed a remarkable fact that mathematics should be so effective in describing the physical world.

How do we get numbers out of physical theories? The kinds of numbers produced by physical theories are just those numbers that we have called computable. They can be obtained by running an algorithm on a universal Turing machine. Indeed much of modern theoretical physics is becoming increasingly computational, to the extent that most Physics Departments, including my own, run courses in Computational Physics. We can approximate any number of physical systems by simulation. Usually, the better the approximation, the longer the program takes to run. Either classical or quantum systems may be simulated to an arbitrary degree of accuracy in principle, however, the time taken to run them may be truly cosmic. This is a particularly acute problem for systems with a number of interacting particles. It does not matter if the problem is quantum or classical. Such complex systems are also a major headache for theory, as well as computation. If the world was classical and not quantum there is a simple trick to get a good simulation in such cases. It is based on accepting that only a statistical description is practical, and using simulated coin-tosses to solve for some approximate physical quantity. However, the world is quantum and thus irreducibly random. That sounds like just the thing for a statistical description based on simulated coin-tosses. However, the randomness in quantum mechanics is not like ordinary coin-toss randomness. To simulate quantum systems by simulated coin-tosses is going to be impossible unless we can generate quantum statistics. For this task we need a new kind of Turing machine, a universal quantum Turing machine.

Suppose we try to simulate the motion of all the particles in a gas as if they were colliding billiard balls. The theory that describes this system in principle is Newton's physics. To make this theory work we need to know the position and velocity of every particle in the gas to a truly astounding level of accuracy. However, in the nineteenth century, physicists realised that to attempt to apply Newton's equations to all the particles in a gas is a hopeless and even useless task. After all, for most purposes we only need to predict a few numbers, such as pressure, temperature and volume. In the late nineteenth century and early twentieth century, a new statistical mechanics was developed to deal with the problem from this perspective. The description was based on computing the *probability* that a particle would occupy a given position with a given velocity. Newton's equations are then replaced by equations that tell us how this probability changes in time. These equations are usually still too complicated to solve by computer. There is a better approach.

Let me take a simple example to illustrate the point. In Chapter 2 I mentioned the phenomenon of Brownian motion, the apparently random motion of pollen particles suspended in water. Einstein explained this phenomenon in terms of the physics of Newton by saying that the motion of the pollen particle resulted from countless collisions between it and the molecules of water. That is the correct physical explanation. Now let us turn to the task of describing it in some detail. We could try and explain the motion by calculating the detailed trajectory of each and

106

every water molecule and the trajectory of the pollen particle, carefully using Newton's equations to describe all the various collisions involved. Needless to say Einstein didn't even attempt such an explanation. Einstein realised that what was called for was a statistical explanation. That is not to say Einstein doubted that Newton's equations could describe the phenomenon, but just that such a description was an unnecessary extravagance. Instead Einstein set out to determine the probability that a pollen particle at a given point in space would, a short time later, be found either in the same place or at some distance away. To do this he would ignore the detailed motion of the water molecules entirely, and use only gross features of the water such as its temperature. The result of Einstein's theory of Brownian motion is an equation that enables us to calculate the probability of finding the pollen particle at any point in space. The description of the water and its interactions with the pollen particle are described by only one number, the so-called diffusion constant. In this explanation, space is the usual continuum of classical physics, as is time. Einstein's equation can be solved by standard mathematical means. But there is another way to solve the problem. We can simulate it on a computer.

The first problem we face is that 'position' in Einstein's equation is a continuous variable, as is 'time'. In this problem we see the fingerprint of infinity. Position and time are infinitely divisible in classical physics. That is no good for a computer. We need to represent both space and time as a finite decimal number. The degree of precision is limited by the size of the computer memory. So straight off we need an approximation for space and time. The way this is usually handled is to divide space up into a grid of little boxes. Instead of giving the position of the pollen particle we simply give the coordinates for the little box in which the particle can be found. Of course our grid is necessarily finite as well, but we make it big enough to ensure that there is very little chance of the pollen particle jiggling its way out of the edges over the time interval of

107

interest. A large three-dimensional array is hard to picture so let us simply use a one-dimensional array, a line of boxes. We have reduced the system of a pollen particle undergoing Brownian motion to the far simpler problem of a particle jumping on a line of little boxes.

We may not know for sure which box the pollen particle starts in. In that case we give a set of probabilities to find the particle in each of the boxes. In classical physics we assume it really is in one or the other of the little boxes. We just don't know which one. Already we have introduced our classical explanation in terms of hidden variables to explain any randomness in observations on the initial position. In a similar way we will divide up time into a series of finite-sized steps. Now the problem can be phrased as follows. Given the set of probabilities to find the particle in the little boxes at one time, find the new probabilities to find the particle in the little boxes at the next time step. Once we have all the probabilities at any time step we can start to calculate interesting things like the average position of the particle at any time. Suppose there are N boxes along the line. We then need to give at least N numbers corresponding to the probability to find the particle in each little box. In the next time step we need other N numbers and these may be quite different from those at the previous time step. In general computing the new probabilities will require in the order of N^2 calculations at each time step, as each of the new probabilities can depend on any of the N probabilities at the previous time step. In some special cases we may be able to get away with a lot less. If N is small this is not too great a problem, but as N becomes large, so that we use a very fine grid, the problem becomes more difficult. If instead we have more than one particle moving on the grid, the problem quickly becomes intractable. For example, suppose we have N particles and we need a grid fine enough to ensure that on average we have only one particle per box. That could be done with roughly N boxes. Now the probability to find a particle in any box may depend on the positions of all the other particles. Just to store the initial probability is going to require of the order of N^N numbers. The number of calculations per step is then going to be roughly N^{2N}. That is getting totally out of hand. For example, if N is only 10, then N^{2N} is 1 followed by twenty zeros. The ultra-computer working at 1 teraflop will take about three years just to compute the first time step!

There is a better way. Suppose instead we start with the pollen particle in a particular box. We choose this box at random, but with odds that reflect the initial uncertainty. For example, we may know absolutely nothing about the position of the particle. In that case, by Laplace's rule we assume here is an equal chance to find it in any box in the grid. In any case we start with the particle in one box, and to define which box only needs a single number. If we had N particles, with one particle per box on average, we would need N numbers to give the initial box of each one. For simplicity, I will assume that the particle always changes its position at each time step and that it does so by moving into one of the immediately neighbouring boxes. The particle can move one step to the left or one step to the right.

As a simple case, suppose that the particle can move to the left or right with equal probability. We can simulate this by tossing a single coin. If the result is heads, move the particle to the left by one box. If the result is tails, move the particle to the right by one box. If we had N particles we need to do only about N calculations per step. This is a lot better than N^{2N}. Of course a long run of coin-tosses will only give us a new configuration for the particles on the grid. To compute averages, we would need to repeat the simulations and take the average over all the runs. However, you don't need to do that many runs to get pretty good convergence to an average. Certainly you don't need to do anything like N^{2N} runs.

The key to solving this problem on a computer is the ability to simulate a coin-toss. This is a relatively easy task. All we need to do is to find some algorithm that produces either a one or a zero with equal probability. How can an algorithm produce a truly random number? Well, it cannot in principle, but it can for all practical purposes. For example, it may turn out that only if you run the algorithm a billion times will you begin to see some pattern in the way zeros or ones turn up. There are a number of such algorithms. Sophisticated computational environments such as *Mathematica* come with built-in random number generators. Solving a physical problem using random numbers to generate a coin-toss is a very well developed area of computational physics. For ease of reference I will refer to all such methods as Monte-Carlo methods, although this name is usually used in a more restricted way. Monte-Carlo methods are very efficient and enable us to solve a problem far more quickly than a head-on assault using Newton's equations.

Now imagine trying to perform this same task in a quantum world. We know that this world is irreducibly random. Isn't that just an ideal place to apply our Monte-Carlo methods? The essential lesson of the quantum theory is that while the universe is irreducibly random it is random in a way that cannot ever be explained by a coin-toss type of randomness. We cannot explain the randomness by proposing hidden properties that turn up at random. The rules of the game are based on Feynman's rule, not the classical probabilistic rules of Laplace. Simulating a random quantum world is going to be fundamentally more difficult than simulating classical randomness. I think the first person to appreciate this point was Richard Feynman himself. Furthermore, not only did he see the problem, he saw the solution—build a quantum computer. In his paper *Simulating Physics with Computers*, Feynman described the problem with breathtaking simplicity, and initiated a line of inquiry that has led us to the realisation that a quantum computer can solve many problems far more efficiently than a classical computer.[5]

Let me now return to the issue of the Church–Turing principle. How well can a computer simulate reality? To explain the content of the Church–Turing principle I want to propose a way to test my imaginary program: *Reality 2001.* The idea is based on a proposal first made by Allen Turing himself to address the question: Can a computer think? The proposal has come to be known as Turing's test. Suppose you have the next iteration of the ultra-computer, the hyper-computer, which computes so fast that we need to invent a new word, the hyperflop, to describe how many floating point operations it can perform per second. One hyperflop is more flops than you can ever imagine. Even better, the machine comes with so much memory that we need a new unit for that too, the hyperbyte. The hyper-computer starts up, quickly crunches through to a couple of hyperflops and starts filling its massive store of hyperbytes with useful things. Now, can the hyper-computer think just like you, me and Stephen Hawking? Suppose the manufacturer claims that the hyper-computer can think? Is this true or just hyper-hype?

Turing suggested there is only one way to find out: If it walks like a duck, if it quacks like a duck, then it is a duck.

In 1950, Turing published *Computing Machinery and Intelligence,* which contained the description of a simple way to test the hyper-computer. Let me put this in more modern terms. Suppose the hyper-computer communicates with the world by simulating a videophone conversation, something like what you get when you run *CU-SeeMe* or equivalent videophone software on your personal computer. When you turn on the hyper-computer, a small window comes up your screen, with the image of a person we will call Alice. Alice looks and sounds as if she is a real person on the other end of the videophone line. To make it more interesting, suppose that there is also another window on your screen that really does correspond to someone, called Bob, on the other end of a videophone line. Now Turing's test is this: Is there any conversation you can hold with both Alice and Bob that would indicate which was the computer? Bob must be truthful and will try and persuade you that he is the real thing. On the other hand, Alice is allowed to lie and is also trying to persuade you that she is the real thing. If after extensive testing you cannot distinguish the real person from the computer, the hyper-computer passes the Turing test. Before you log-off, you arrange a date with Alice.

One way to run the Turing test on the hyper-computer would be to ask it to play a game of chess with Gary Kasparov. Let us assume that Bob is also a grandmaster. A few years ago Gary Kasparov would have had little difficulty in flunking the computer, but that has certainly changed. Recently IBM's DEEP BLUE beat Gary Kasparov. IBM Research's DEEP BLUE project team, Feng-Hsiung Hsu, Murray Campbell, A. Joseph Hoane Jr, Gershon Brody and Chung-Jen Tan, saw this problem as a good template for similar complex problems involving search routines. DEEP BLUE computer is a 32-node IBM RS/6000 SP-2 high performance computer. It contains a total of 256 processors working in tandem. The machine can calculate 50 to 100 billion positions in three minutes, the time allowed for moves in major tournaments. Computers can play chess at grandmaster level. Or do they? What if Gary Kasparov was not in fact playing at all. Suppose he had a small speaker in his ear connected to the hyper-computer. As each move is made by DEEP BLUE, the hyper-computer tells Gary what move to make next. It is not Kasparov versus DEEP BLUE but the hyper-computer versus DEEP BLUE. Would a grandmaster watching this game ever know the difference?

I am going to propose an alternative to the Turing test for testing the Church–Turing principle. Here is the story. Suppose in some well-resourced University Physics Department a graduate student is locked away in her laboratory conducting experiments on aspects of reality. The student can draw on a finite but arbitrary supply of resources: equipment, computers, paper and of course a vast amount of coffee. Her adviser, however, is forever imprisoned in his office writing grant applications to keep his students well supplied. The only way student and adviser communicate is through e-mail. Each morning the student logs on to receive an encoded binary file which contains the instructions for the day's experiments. Each day brings a new but finite set of instructions for conducting the experiment. When decoded these instructions are a mixture of mathematical and natural language statements that the student can interpret. There is no quantum magic in the way these instructions are given. They could all just as well be given verbally, if only the adviser could ever get out of his office. The student receives the instructions, orders the equipment for the day and conducts the experiment. The result of the experiment is a set of numbers representing the results of the measurements, possibly including some binary encoded postscript files for graphical output, and also including a set of plain language statements describing what the student actually did, that is, how the equipment was assembled, how the data was taken and analysed and any problems that remain unresolved. All this is sent off via e-mail to her supervisor, who decodes the binary files and takes them home to think about after dinner. The next day the chain of events repeats. The situation is depicted in Figure 4.3.

One day while attending a lecture on the Church–Turing principle, she gets a great idea. She realises that as far as her supervisor is concerned, she may as well be a computer. After all he 'inputs' binary data as instructions for the day and she 'outputs' binary data representing the results. All the supervisor sees are sets of numbers, a few pictures and a bunch of sentences. One week, while her adviser is off presenting a paper at a conference, she buys a copy of *Reality 2001*, and gets busy programming the lab's hyper-computer to receive the supervisor's instructions and run a simulation of the proposed experiment. She watches for a few days until convinced that everything is working fine. When an angry supervisor never appears at her door she decides that the hyper-computer can take care of itself and goes to the beach for the summer. Will her supervisor ever know the difference?

113

Figure 4.3 A test of the Church–Turing principle. Bob asks Alice to perform experiments by issuing instructions through e-mail. Alice reports back her results through e-mail. Can Alice simply feed Bob's messages to a hyper-computer and run the experiments in 'simulation'? Is there any experiment Bob can specify that would enable him to find the experiments were virtual?

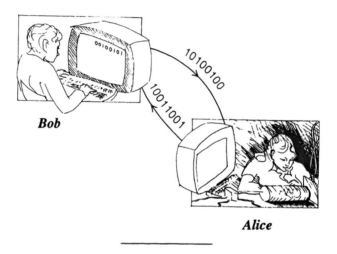

Bob

Alice

According to the Church–Turing principle the answer is no, so long as the hyper-computer is a quantum computer. If it were not, the supervisor would soon discover all was not as it seemed. If the hyper-computer was only a classical computer and the supervisor asked for some experiment involving quantum entanglement of many particles, the game would be up. Hyper-computer has been programmed well enough to know that it is being asked a quantum question and does not even try to answer it using a Monte-Carlo method. Instead it tries to calculate all the relevant probability amplitudes in detail. This would be like a classical computer trying to calculate the probabilities to describe the interactions of many particles. It is going to take a very long time, a time that rises exponentially with the number of entangled particles involved. The supervisor would begin to suspect that his student was asleep on the job when no results ever came back. If instead the hyper-computer tried to take a short cut by using classical Monte-Carlo methods, the supervisor would immediately see that the results disagreed with the quantum theory. However, if hyper-computer was a quantum computer, a device that exploited quantum entanglement directly, the supervisor would go on entering instructions and writing papers based on the results, as if real experiments were being done, and never know the difference.

Just what do I mean by a quantum computer? In the next chapter this question will be answered, with some surprising consequences for physics, but we can make some preliminary observations. To begin with, a quantum computer must be a real computer, that is to say, it must be capable of universal computation in the same sense as the universal Turing machine. As a universal Turing machine it must therefore be constrained by the logic of the halting theorem. A quantum computer cannot compute uncomputable functions. No quantum computer can solve the halting problem.

However, a quantum computer can do things a classical computer cannot do. An arbitrarily accurate simulation of a quantum physical system, in a time that does not grow exponentially with the size of the problem, is one example. This last observation immediately suggests that some computations might be much more efficient on a quantum computer. Efficiency is not the same as computability. A function may be computable, but take so long to perform on a universal computer that for all practical purposes it cannot be computed in a reasonable time. We saw one example of a difficult simulation in trying to solve the problem of N particles moving randomly on a grid of cells. In that case a direct head-on assault to determine all the required probabilities for all possible arrangements of the particles on the grid would take a time that scaled as N^N, that is, exponential in the problem size variable, N. We could get a more efficient approach by directly simulating coin-tosses as part of the algorithm itself, a Monte-Carlo algorithm. The possibility of more efficient algorithms which exploit quantum physics has since been realised, as we shall see in the next chapter.

A computer is a physical device, a machine, that is constructed to realise a computation, a mathematical concept, as it runs. A quantum computer is likewise a physical system, but one that runs according to quantum physics not classical physics, to realise a computation. What can we change in a Turing machine model to make it look like a quantum system? First, the tape could be a quantum system. We must be able to prepare a physical system in a quantum state that corresponds to the data and program for the desired computation. To achieve this we can place a qubit instead of a bit in each cell on the tape. Second, the head could be a quantum system, with distinct physical states described by quantum physics, not classical physics. Finally, the dynamics of the machine, how it moves in response to the states in each cell on the tape, could be determined by quantum physics instead of classical physics. There is one other thing that is required: we must be able to make measurements on the physical system representing the output tape. At the end of the day we need actual bits not qubits. This last requirement takes us back to the beginning of the quantum theory and a rather subtle principle.

Recall the story of the clever student and the quantum hyper-computer which was able to simulate quantum physics so well that no-one could tell the difference between a lab full of equipment and a simulation of a lab full of equipment. Even though a quantum system can be simulated, the end product is a set of measurement results which are 'written' down in hard bits of data. The supervisor does get a real message which he can read, as if it was a truly classical state. This is a very important and deep issue that was repeatedly emphasised by Neils Bohr, one of the founding fathers of the quantum theory. Bohr said in

116

essence that no matter how weird it gets, the results of all experiments must be communicated in classical terms: '. . . however far the phenomenon transcend the scope of classical physical explanation, the account of all evidence must be expressed in classical terms'.[6]

Indeed it is not only the results of the experiment that must be in classical terms. The instructions to the student are also real bits, representing real words and numbers drawn from our classical vocabulary. The measurement results must always be conditioned on the input instructions. We need to know both the code of instructions and the binary code of measurement outcomes to have an exhaustive understanding of what was done. Despite this, we can discover a new aspect of reality, the quantum. It is precisely the unexpected correlations between the input binary code for the hyper-computer and the output binary code, that leads us to conclude that a quantum hyper-computer is a physical device dancing to the tune of the quantum principle.

I am thus led to a slight reformulation of the Church–Turing principle.

> The results of all finitely describable physical measurement systems can be perfectly simulated by a universal quantum computer operating by finite means, and the resulting measurement records are terminating.

By 'finitely describable physical measurement systems' I mean that the instructions for building and operating a measurement device, including both the system and apparatus, must be able to be represented by a finite code. In other words, the instructions to the unfortunate graduate student of our parable must be finite and not run on forever. A perfect simulation of the measurement results means that the numbers produced by the simulation must be precisely the same as those obtained from a real measurement. In some cases these numbers will be random. In that case further statistical analysis of the numbers will reveal probabilities indistinguishable from what would be

117

obtained in a real experiment. Finally, the requirement that the measurement records for a given simulation be terminating, means that all simulated measurements must come to an end some time. A measurement scheme is not to go on producing results forever. The form of the Church–Turing principle I have just described might well be called the Bohr version of the Church–Turing principle. Is it correct? I think we should postpone judgement until we have looked a little more closely at just what a quantum computer is.

CHAPTER FIVE

QUANTUM SOFTWARE

Two powerful intellectual tracks converge on the technology horizon of quantum computing. The first track begins with Richard Feynman's question. Can quantum reality be simulated? The answer to this question will resolve the issue of whether a perfect simulation of reality is possible in principle, the Church–Turing principle. In 1985, David Deutsch of the University of Oxford demonstrated that an affirmative answer to Feynman's question would have important practical consequences for computation. By explicitly utilising Feynman's rule for combining probability amplitudes, a quantum computer could in principle solve problems that are intractable on a classical computer. I shall return to this track later. For now, I want to follow the second track leading to quantum computing. This track originates in a technological imperative: to build ever more powerful computing machines on an ever diminishing scale.

Quantum technology

The massive increase in computational power represented by contemporary digital computers is a direct result of the successful drive to build such devices on smaller and smaller scales. Computers are machines, and as such are constrained by the laws of physics. Is it possible to continue to make devices smaller and smaller? Are there physical limits to this process of diminution? More generally, are there phys-

119

ical principles that limit the power of computational machines? In the nineteenth century, machines were built to process energy, to convert it from one form to another. When physicists turned their attention to the processes involved, fundamental new laws were discovered to constrain the action of engines. These laws are now embodied in the subject of thermodynamics. The first law of thermodynamics captures a conservation principle: energy can only be converted from one form to another, but the sum total of energy in a closed system remains the same. The second law of thermodynamics constrains how efficiently energy can be converted into useful work. These two laws together imply that a perpetual motion machine is an impossibility. That may seem obvious to you, but it was not at all obvious last century. Even today, the patents offices around the world continue to receive new inventions that promise to run forever without running out of 'steam'.

Computers, like all machines, are constrained by the laws of thermodynamics, but are not built primarily to lift bricks or drive trains. Computers process information, while engines process energy. We have become quite accustomed to thinking of energy as a kind of 'stuff' that can be bought and sold like any commodity. This familiarity has arisen over nearly 200 years of dealing with machines. In the closing days of the twentieth century, we are becoming adjusted to regarding information in a similar way, as a commodity with considerable economic gravitas. Might there be physical laws that constrain our ability to process information? If there are, such laws will have considerable consequences for the Church–Turing principle.

In the 1960s, Rolf Landauer of IBM Thomas J. Watson Research Laboratories began to investigate such issues. Computers consume energy and produce heat, just like an engine.[1] Getting rid of this heat is a real problem for small devices. But is it obvious that computer chips must produce heat? Heat is a form of energy, but is usually not the desired outcome for an engine designed to do useful work, such as getting you to work each morning. The laws of thermodynamics require that heat be produced by all engines that do useful work. How much heat is produced for a given amount of useful work is a measure of the efficiency of the engine. Are computing chips also constrained by such thermodynamic principles? Landauer's answer is yes, but only if we continue to build them the way we currently do. His discovery, known as Landauer's principle, says that heat must be dissipated when information is erased.

Landauer traced the problem, rather surprisingly, to the kind of logic we have built our computing machines to perform. There are a small number of logical operations, operations on binary numbers, that are essential for general purpose computation. You can't get by with less. Just which ones you decide to call fundamental is somewhat arbitrary. One such operation is NOT. This just takes a bit and inverts it, so that a 1 goes to 0 and a 0 goes to 1. When implemented in circuitry the NOT operation has one input and one output. We refer to such a physical implementation of a logical operation as a gate. Another logical operation is AND. This is a function of two qubits. If the two qubits are both 1, the result of the operation is 1, otherwise the result is 0. When the AND gate is implemented in circuitry it will have two inputs and one output. Simple diagrams are used to describe these gates (Figure 5.1).

There is a fundamental difference between the NOT and AND operations. The first is reversible, while the second is not. That is to say, if you know the result of a NOT operation you can infer the input. However, if you know only that the result of the AND operation is 0, you cannot be sure of the input, as three distinguishable alternatives (00, 10, 01) all lead to the same output. This convergence of distinguishable alternatives is reflected in the diagram of the gate: two inputs and only one output (Figure 5.1). The AND gate necessarily destroys information.

Landauer's principle states that whenever a bit of information is erased, a small amount of energy must be given up as heat (roughly equivalent to the kinetic energy

Figure 5.1 These symbols represent logic gates in classical digital electronics. The top symbol represents a NOT gate and the bottom an AND gate. Note that the AND gate has one less output wire than input wires. This means it must be implemented as an irreversible physical system.

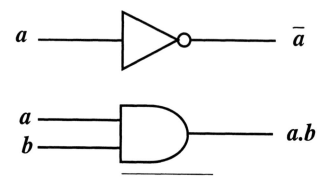

of an air molecule at room temperature). How do we erase information? Return to the example of a coin-toss. If two sides of the coin are distinguishable we gain one bit of information when we learn the result of a single toss. If both sides of the coin were the same we would gain no information upon tossing the coin. Thus to erase information we must arrange for two previously distinguishable alternatives to become indistinguishable. That is exactly what happens in an AND gate. The AND gate implements a logically irreversible operation. It is not possible to invert the operation and recover the input in general. A logically irreversible operation must erase information and any physical realisation of that operation will pay the price of Landauer's principle. Any computational device built out of AND gates is necessarily going to get hot. Similar constraints would arise for Turing machines. At the end of a computation, many Turing machines leave only the answer on the tape, having erased the input data. At each step the Turing machine may not be reversible. It can only be reversible if the binary string on the tape after each step has one and only one predecessor binary string.

If a computer is built out of irreversible logic gates it will necessarily waste energy as heat. But is it obvious that we must use irreversible operations to do computations? After all, the NOT operation is completely reversible. This question was answered by Charles H. Bennett in the early 1970s, working in the same IBM research laboratory as Landauer. Bennett proved that given any Turing machine, there is a reversible Turing machine which performs the same computation, and that furthermore the reversible Turing machine will not require many more steps to solve the problem than the original irreversible machine. A reversible Turing machine can be just as efficient as an irreversible one. Bennett's result means that we do not need to erase information to do universal computation. Landauer's thermodynamic cost can be avoided entirely. Computation is cool, in principle.

Bennett constructed his reversible Turing machine using pretty much the same kind of classical intuition as Turing. However, the only truly fundamental Turing machine must be quantum mechanical, and it is natural to ask if quantum mechanics constrains reversible Turing machines in some way. This issue was taken up by Paul Benioff of Argonne National Laboratory in the late 1970s. Benioff proposed a full quantum version of a reversible Turing machine. Benioff's machines exactly mimic the classical machines. They do not exploit the fundamentally new property of quantum mechanics embodied in Feynman's rule, which enables us to combine probability amplitudes for different indistinguishable computational alternatives.

A reversible quantum Turing machine will not dissipate any energy into the environment. In this sense it is completely isolated from the outside world. That is not a very useful computer. Recall the discussion, in the previous chapter, of the student who decides to do her PhD research entirely in simulation. Such a universal reality simulator is not much use if it does not talk to the outside world. It must accept instructions and produce measurement results, both of which are part of the real classical world outside the computer. The way around this problem is by using some part of the quantum computer as a flag to signal the end of the computation. The issue of halting needs to be revised a bit for a reversible Turing machine. A Turing machine is said to halt if two successive configurations are the same. But that is an irreversible process. To fix this we can arrange for the machine to signal the end of a computation by setting some particular cell on the tape to 1 when the computation is over. This special tape cell is called the flag. The machine runs reversibly and without interacting with the external world. After some time, however, it sets a flag to 1, indicating the computation is finished and the answer can be collected from the machine's memory. We can then check to see if the flag is up or down after waiting some suitable time. Alternatively we can forget about the flag and just open the machine up to the outside world from time to time to see if the computation is done. If it is not done, we just start it over again. Either way, the key issue is that while the computation is actually running, the physical system implementing it must remain completely isolated from the external world. Such dynamics is called *unitary*.

Unitary dynamics is precisely what makes quantum computation at once so powerful, but so difficult to implement. While the computation is running, we can never know what physical states the machine runs through. There may be a huge number of possible histories for the operation of the machine that lead to the result we finally see when we connect the device to the outside world. There may be many different ways to achieve the final result, none of which we can distinguish. All of these different histories represent different computational paths, but we can never know which particular computational path the machine has followed. That is exactly the kind of thing to which Feynman's rule applies. It was this possibility that intrigued Feynman and was made explicit by Deutsch.

Reversible gates

Every digital computer is a realisation of a universal Turing machine, but they are not built out of long paper tapes and moving heads. Digital computers are built by finding physical devices to implement logical gates. It can be done by a combination of NOT and AND gates arranged in suitable sub-arrays to form more complicated gates. A combination of such gates can be completely equivalent to a universal Turing machine. In every case, however, the gates used implement logically irreversible functions and thus are physically irreversible as well. Current proposals to realise quantum computers are also based on gate arrays, rather than directly reproducing a Turing machine, but the gates themselves must be physically reversible if the processing is to be unitary.

In the early 1980s a number of proposals for universal reversible logic gates were published. Of these, the Fredkin gate and the Toffoli gate have received the most attention for quantum computation. The Fredkin gate and the Toffoli gate, if they are to be reversible, must have the same number of inputs as outputs. It turns out that universal computation requires a fundamental reversible gate with at least three inputs (and of course three outputs). All the functions of standard logic can be implemented by a suitable arrangement of such gates.

The best way to describe a reversible gate is to give a table listing the logical value for each output corresponding to a given input. As there are three inputs (and three outputs), all possible inputs can be represented by three binary symbols: a three-bit input. Likewise for the output. The Toffoli gate can be represented by Table 5.1 and the Fredkin gate is represented by Table 5.2.

In a Toffoli gate, two input bits A and B control the state of a third 'target' bit C. The control bits do not change between input and output. The target bit will change its state only when the two control bits are both set to 1. In a Fredkin gate there is one control bit and two target bits. In the table, the control bit is C and the target bits

125

Table 5.1 A Toffoli gate   *A,B=1,1*   *C changes*

INPUT			OUTPUT		
A	B	C *Target*	A	B	C *Target*
0	0	0	0	0	0
0	1	0	0	1	0
1	0	0	1	0	0
1	1	(0)	1	1	(1)
0	0	1	0	0	1
0	1	1	0	1	1
1	0	1	1	0	1
1	1	(1)	1	1	(0)

The logical table for a Toffoli gate, a universal gate for reversible computation. We may regard the C bit as a 'target' bit which changes its state only when the two 'control bits', A and B, are both set to 1. In all other cases C does not change its state. In every case the state of the control bits A and B never change.

Table 5.2 A Fredkin gate   *A, B interchange*   *C fixed*

INPUT			OUTPUT		
A	B	C	A	B	C
0	0	0	0	0	0
0	1	0	0	1	0
1	0	0	1	0	0
1	1	0	1	1	0
0	0	1	0	0	1
0	1	[1]	1	0	[1]
1	0	[1]	0	1	[1]
1	1	1	1	1	1

The logical table for a Fredkin gate, a universal gate for reversible computation. We may regard the C bit as a 'control' bit which never changes its state. However, if the bit on the control is 1, the bits on the other two lines are interchanged.

126

are A and B. The control bit does not change. The target bits swap only when the bit on the control is set to 1, otherwise they are unchanged. Which of these one chooses to implement depends to some extent on the physical system used. For example, the Fredkin gate has often been used for photon based gates where a 1 represents a photon and a 0 simply represents no photon. In that case the number of ones cannot change as the number of photons cannot change (absorption is not allowed for reversible gates). The Fredkin gate preserves the number of ones between input and output, while the Toffoli gate does not. However, both gates are logically reversible and thus in principle can be realised as physically reversible devices. Each type of gate is universal in the sense that all other logical functions can be implemented by a suitable arrangement of either gate. An example of how to build a NOT gate and an AND gate from a Fredkin gate is shown in Figure 5.2. In both cases there are more outputs than are required for the function. These outputs, called garbage bits, are a necessary consequence of reversible logic.

In reversible computation the number of garbage bits increases dramatically, placing big overheads on the storage capacity of the device. These bits cannot be irreversibly erased as that would require that we pay the Landauer price, and render the operation of the array non-unitary. The device could not realise the power of quantum computation inherent in Feynman's rule. Bennett realised that it is not necessary to erase the garbage bits. They may be reversibly reset to some known fixed value. The inevitability of garbage bits means that we must allow for many more input bits than are needed for irreversible computation. While some of the garbage bits may be entirely internal to the machine, a certain minimum number of additional input bits will be required to take account of these garbage bits.

At first sight it looks to be an impossible task to say in general just how many garbage bits will be needed for a given computation. However, it is quite easy. The simplest way to ensure a computation is reversible is to arrange for

Figure 5.2 The Fredkin gate may be made to operate as a NOT gate or an AND gate by presetting particular inputs. Notice that in all cases there are more inputs and outputs than are strictly required to realise these gates in an irreversible scheme. The unwanted output bits on the lower two lines are referred to as garbage bits.

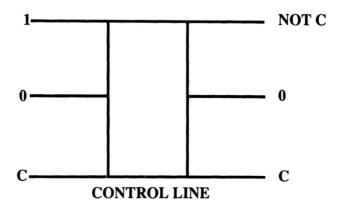

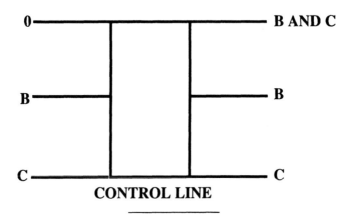

the input to be reproduced, together with the output at the end of the computation. For example, suppose the computation sets out to add three and five. If the output, eight, is produced and the input numbers are erased in the process, we cannot uniquely invert the operation. However, if we set up the algorithm in such a way that both the output number and the two input numbers (one would do) are present at the end of the computation, the operation is clearly reversible as no information has been lost. Bennett realised that a similar idea could be used in a universal Turing machine. At the start of the calculation the tape contains a binary string coding for all the input data, together with a number of blank cells set aside to eventually contain the final result. The computation produces the final result by writing the answer to these blank cells but does not overwrite the input cells.

To do this, we run the machine to produce the output, reversibly copy the output to a preset register, then run the computation backwards to return the input and a set of preset registers. The idea is illustrated in Figure 5.3. Needless to say a huge number of gate operations may be required inside the box 'computer' of Figure 5.3 to ensure any necessary garbage bits are reversibly reset to some initial state.

The work of Landauer and Bennett laid the foundations for a physical theory of computation. Landauer showed us that logical irreversibility would necessarily entail physical irreversibility in the machines designed to implement the logic. Bennett showed us how to avoid physically irreversible computers by using only logically reversible operations. By the end of the 1980s it was clear that reversible computation was not only possible, but desirable, even if only in an approximate form. If you don't need to waste energy at every step then why do it? Getting the equivalent computational power for a lower energy cost would reduce the voracious appetites of laptops for battery power. Technological benefits aside, the idea of reversible computation cleared the way for a fully quantum computer operating

Figure 5.3 A scheme for reversible computation. An initial output, together with a number of preset registers, is processed in the computer. The ouptut is then reversibly copied to one of the preset registers. The calculation is then run through a time reversed computer. At the end we have the input and the output. The computation is thus reversible. All garbage bits are reversibly reset to 0.

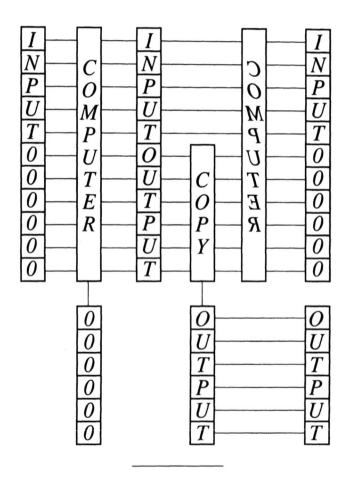

entirely by unitary processes, and in consequence capable of fully exploiting Feynman's rule to achieve computational goals.

Quantum data storage

The time has come to look more deeply into how a quantum computer might be built. The first element of a computer that we need to think about is data representation. We will then need to consider how to build quantum gates for processing the data. I will continue to assume that all data is represented by a binary code of only two symbols 1 and 0. Each symbol represents a single bit of information. In a quantum world binary systems, qubits, abound. We have seen a number of examples: magnetic orientation, the behaviour of a photon at a beam-splitter and photon polarisation. In these examples, the result of each measurement is one of two mutually exclusive events which we can code as a 1 or a 0. When we make a measurement on a qubit the result may be certain, or it may be as uncertain as a coin-toss. It all comes down to the pre-measurement amplitudes for each of the two mutually exclusive outcomes. But we know that these systems are very different from a coin-toss. The probabilities are determined at a fundamental level by probability amplitudes which are combined according to Feynman's rule. This leads to some highly counter-intuitive behaviour, particularly for entangled systems. Thinking about such cases we were led to abandon Einstein's belief that the uncertainty in experimental results was due to lack of knowledge of some hidden variable. There can never be some randomly distributed 'gene' that predetermines the outcome for a particular measurement. To capture this fundamental difference in the behaviour of binary quantum systems we use the word *qubit* rather than bit to describe the fundamental quantum event waiting to be realised by an act of observation. Can we use qubits rather than bits to store information? The answer is yes, with some very surprising consequences.

131

Qubits are a novel way of representing information. I discussed the concept of a qubit in Chapter 1, but let us revisit the idea. I will take as my example of a qubit the direction of a single photon after it encounters a beam-splitter (see Figure 5.4).

The incoming photon is described completely by giving its direction of travel.[2] In the figure there are two possible paths for the incoming photon and two possible paths for the outgoing photon. I will refer to each of these paths as U and D, according to whether the path is heading up towards the top of the page or down towards the bottom of the page. In Figure 5.4 the possible paths are shown as dotted lines. The fact that there are the same number of input directions as output directions is very important. It indicates that no information is lost at a perfect beam-splitter. A beam-splitter must be a unitary device, which means it can be an ingredient in a quantum gate.

In Figure 5.4, only a single photon is incident from the D-direction. Suppose there are two photon detectors after the beam-splitter, one placed in the path of the U photon and the other placed in the path of the D photon. The photon is equally likely to be detected at either photon detector. This is because the single photon can be either transmitted into the output path D, or reflected into the output path U by the beam-splitter. At the quantum level each of these detection events is determined by a probability amplitude.

Now suppose the photon was incident in the U-direction. This photon will be detected by one or the other detector with equal probability, just as for the case of an incoming D photon. Does this mean that the same probability amplitudes determine the detection probability for either a U or a D incoming photon? The answer is no. Even though the probabilities for detection are the same in both cases, the two situations are physically distinguishable. The source of the incoming photon may contain a record of the direction of travel. Perhaps the sources are two independent lasers with a movable shutter covering the

Figure 5.4 A beam-splitter for photons mixes two distinguishable photon paths. One path, U, is directed towards the upper part of the page, while the other path, D, is directed down the page. An incoming photon in the D-direction is randomly directed to detectors placed in the outgoing U- or D-directions, with equal probability. Two independent lasers are used as the photon sources. A shutter allows either a U- or a D-directed photon to be sent towards the beam-splitter.

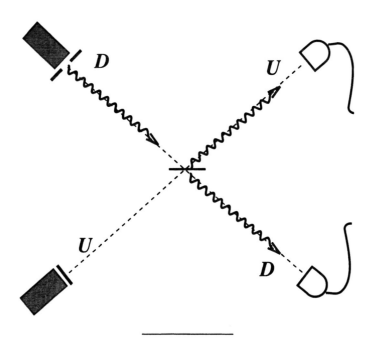

output of each one. Whether we send a D photon or a U photon depends on which shutter is open, and we can always keep a record of this. There are two distinct physical situations at the input and there must be two distinct physical situations at the output, despite that fact that the photon detectors do not distinguish them. We cannot have a merging of two distinct physical states into one for a unitary device. To see this more clearly, suppose we run time backwards just after a photon encounters a beam-splitter. Now we have a single photon heading towards the beam-splitter, from either the U-direction or the D-direction. After the encounter we have two distinct physical situations. Either the photon heads back towards one source or the other. Each of these are distinct physical states. If the physical situation is to be reversible in time, it must be the case that there are two physically distinct states for the photons to the right of the beam-splitter (see Figure 5.5). There must be two physically different states, in which the path of a single photon to the right of the beam-splitter is uncertain. This physical distinguishability must be reflected in the probability amplitudes we assign to the outgoing photons. It means that the probability amplitudes for reflection and transmission for a D photon are *not* the same as those for a U photon, despite the fact that the probabilities for detection are the same.

There are a number of ways to assign the probability amplitude for reflection and transmission of the D photon and the U photon. The subject of quantum optics can be used to calculate these amplitudes exactly. What we get depends on the kind of beam-splitter we are using. Here is one way to do it. The pair of amplitudes for an incoming D photon, after the beam-splitter, are chosen to be $(\frac{1}{\sqrt{2}}, \frac{1}{\sqrt{2}})$, and the pair of amplitudes for the U photon, after the beam-splitter, are $(\frac{1}{\sqrt{2}}, -\frac{1}{\sqrt{2}})$. The first element in each pair gives the probability amplitude for transmission (T) while the second gives the probability amplitude for reflection (R). Note that one of the amplitudes is negative, the feature of probability amplitude that distinguishes it from real probabilities. Once we have these amplitudes it is easy to determine the probability for detection at each of the detectors. If, for example, we have an incoming U photon, it will be reflected towards the D-detector with probability amplitude $-\frac{1}{\sqrt{2}}$, and thus will be detected there with probability $\frac{1}{2}$.

Regardless of whether we use an incoming D photon

Figure 5.5 There are two physically distinguishable ways a photon can encounter a beam-splitter. In (a) a photon approaches a beam-splitter from the D-direction, while in (b) a photon approaches a beam-splitter from the U-direction. There must be two distinct physical states after the beam-splitter, if the device is to be physically reversible. This means there are two distinct sets of probability amplitudes after the beam-splitter, as indicated by the two different directions for the probability amplitude arrows. The effect of this is to make the probability amplitude for reflection or transmission of a D photon different from the corresponding amplitudes for a U photon. In the geometry of the figure the effect of the beam-splitter is simply to rotate the amplitudes for either a U-directed photon or a D-directed photon by 45 degrees. This kind of rotation of an amplitude is characteristic of a unitary (or reversible) process.

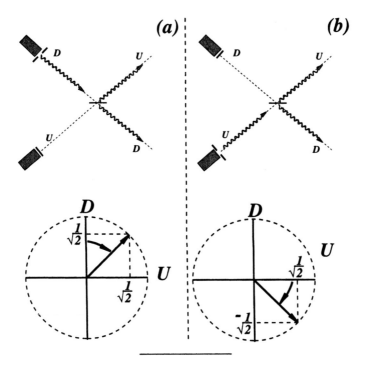

or a U photon, the output state of the photon is described by a probability amplitude pair, one amplitude for reflection and one amplitude for transmission. When we measure which of the two mutually exclusive possibilities is realised, we always find one or the other with equal probability. We get a single bit of information. However, as the source of our information is encoded in probability amplitudes, we know our information is encoded as a qubit and not a simple bit.

In the example, the incoming photon is either a U or a D photon and in principle we can distinguish which is which. We could make the choice of the source random, by tossing a coin to determine which laser shutter to open. The incoming photon could then code an ordinary classical bit of information. The beam-splitter thus provides a way of turning classical bits into qubits. Now there is something rather remarkable hiding here. The incoming photon can be either a U or a D photon, and each is either reflected or transmitted with equal probability. However, the probability amplitudes are not necessarily all equal. We only require that the square of the amplitudes be equal. We thus need in general *four* numbers to describe what happens at the beam-splitter: two probability amplitudes for the U photon and two probability amplitudes for the D photon.[3] If the beam-splitter did not work at the level of qubits, but instead simply performed a coin-toss to determine if a photon was transmitted or reflected, we need only give two numbers, the probability for reflection and the probability for transmission. An incoming U photon and a D photon both have the same probability for reflection and transmission.

The state of a single photon after a beam-splitter is uncertain. We do not know which path, U or D, it follows. However, this is not a coin-toss kind of uncertainty. A photon incident from the U-direction can take either the U-path or the D-path, but we cannot know which, until we make a measurement. According to Feynman's rule, if we cannot distinguish which path the photon took after the beam-splitter, we must assign a probability amplitude to each possibility. To capture the essential difference between the quantum description of a photon at a beam-splitter and a classical coin-toss, we say that the photon is in a *superposition state* after the beam-splitter. In fact, for a beam-splitter the state is an equal superposition of the two output paths. The idea of a superposition state is really nothing more than a restatement of Feynman's rule. A photon in an equal superposition state of a U-path and D-path is a state in which the path is equally uncertain. Furthermore, there are two physically distinct kinds of superposition states after the beam-splitter. One corresponds to the output state of an incoming U photon and the other to the output state of an incoming D photon.

If we measure which path the photon took, we find that it takes a U-path or D-path with equal probability. It looks like a classical coin-toss at the beam-splitter, but it isn't. Suppose we remove the detectors after the beam-splitter, and replace them with two perfectly reflecting mirrors that redirect the paths back to another beam-splitter, as in Figure 5.6.

If the first beam-splitter were a classical coin-toss, so is the second beam-splitter. In that case we would find the probability for detecting the photon in the U-direction after it finally leaves the device is $\frac{1}{2}$, as is the probability for detecting the photon in the D-direction. But the state of the photon after the first beam-splitter is a superposition state of a D-going photon and a U-going photon. There is no way we can know which possibility is actually the case. At the next beam-splitter, the output state of a photon is again determined by probability amplitudes, and we can combine the various amplitudes for different photon histories quite easily. For example, the amplitude that the incoming D photon is transmitted at the first beam-splitter is $\frac{1}{\sqrt{2}}$. This photon then encounters a perfectly reflecting mirror which turns it in the U-direction. Now it encounters the next beam-splitter. The amplitude that this photon is reflected at the second beam-splitter is $-\frac{1}{\sqrt{2}}$. Finally it will

137

Figure 5.6 An incoming D photon is sent into a dual arrangement of beam-splitters. This photon will emerge with certainty in the U-direction. This is easy to understand in terms of the probability amplitudes rotations that take place at a beam-splitter (see Figure 5.5). Each beam-splitter rotates a probability amplitude by 45 degrees. So the two beam-splitters effect a rotation of 90 degrees overall. Likewise an incoming U photon will emerge with certainty in the D-direction. If we encode the U-direction as a 1 and the D-direction as a 0, the device implements a NOT gate.

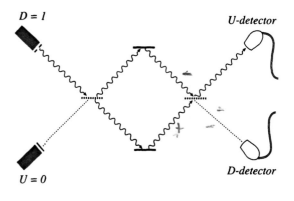

head off in the D-direction. The amplitude for a photon incident in the D-direction, to first be transmitted and then to emerge in the D-direction, is given by the product of these two amplitudes, that is, $-\frac{1}{2}$. However, there is another way in which a photon incident in the D-direction can end up heading in the D-direction. It can be reflected at the first beam-splitter, and transmitted at the second beam-splitter. The total amplitude for this is $\frac{1}{2}$. No minus sign! But we cannot distinguish these two different histories, so according to Feynman's rule we must add the amplitudes together. When we do that we see that they cancel exactly. The amplitude for an incoming D photon to follow the history TR is $-\frac{1}{2}$, but the amplitude for the photon to

138

follow the history RT is $\frac{1}{2}$. An incoming D photon that is detected in the D-direction can follow RT or TR, so we add the respective amplitudes to get 0. The probability amplitude for the photon to be detected in the D-direction is 0, so the probability for detection in this direction is also 0. No photon will ever emerge in the D-direction! The result of the experiment is completely certain. A photon incident in the D-direction will emerge with certainty in the U-direction. That is as far from the classical coin-toss prediction as it is possible to get. When the experiment is done we do indeed see that no photon ever emerges in the D-direction. Probability amplitudes, not probabilities, describe the state of a photon at a beam-splitter. The state of a single photon after a beam-splitter is indeed a super-position state, not simply the unknown result of a coin-toss. A sequence of coin-tosses is still a coin-toss, but a sequence of qubit-tosses can turn irreducible uncertainty into certainty.

We can toss a coin to decide whether to send a photon in the U-direction or the D-direction towards a beam-splitter, but after the beam-splitter the photon is in a superposition state of the D-direction and the U-direction. The input state is described by a single bit of information. We could, for example, record which kind of photon we sent towards the beam-splitter by a single binary number, 1 for a U-directed photon and a 0 for a D-directed photon. The incoming photon represents a single bit, but the outgoing photon represents a qubit, not a bit, as it is in a quantum superposition state. Of course that does not matter much if we immediately measure which path the output photon is in. That gives equal probability for both U and D. However, we can continue to manipulate the probability amplitudes as in Figure 5.6. That is the key idea leading to a quantum computer. We can postpone the final readout to the very end of the processing. The final result must be one of two mutually exclusive possibilities, but while the amplitudes themselves are being processed, we have super-position states of mutually exclusive possibilities, that is, we

139

have qubits. If we can find some way to encode an input binary string in qubits rather than bits, we can manipulate them directly as qubits, before finally making measurements and returning qubits to ordinary bits of information.

To see the enormous power made possible by processing qubits rather than bits, let us look at a way to prepare an input 'tape' to a Turing machine in a superposition of all possible binary strings of length N. First we need some way to encode our bits and qubits in physical states. I will again use the two mutually exclusive possibilities at a beam-splitter, that is, either reflection or transmission. Suppose we have N beam-splitters arranged as in Figure 5.7.

For each beam-splitter there are two possible input paths, corresponding to a photon heading in the U-direction and a photon heading in the D-direction. This gives N different paths heading down and N different paths heading up. At each beam-splitter we have two lasers with a shutter at the output. This enables us to choose which direction to send the photon towards each beam-splitter. I will code a D-directed photon as a 1 and a U-directed photon as a 0. To specify the input state to the entire array of beam-splitters we need to give a binary string arranged in an order corresponding to the order of beam-splitters. For example, if the binary string consists of N 0s, we open all the shutters on the U-directed lasers and send in a single photon in the U-direction to each of the N beam-splitters. Or we could toss a coin at each laser to decide which shutter to open. In that case the binary string that codes for the input would be a random string of ones and zeros drawn from 2^N equally likely binary strings of length N. Note that the input to the entire array is specified by an ordinary classical binary string. The input to a quantum computer is at the very start a classical code. Recall the parable of the student in the last chapter. The student received instructions from her adviser as an ordinary binary encrypted file via e-mail. The student may then go on to either actually do the experiment or program her quantum

Figure 5.7 A set of three independent qubits, all initially set to 1, are prepared in an equal superposition of a 1 state and a 0 state. The net effect is to prepare the input to the quantum computer as an equal superposition of all possible binary strings of length 3. The quantum computer can continue to process all these strings in parallel.

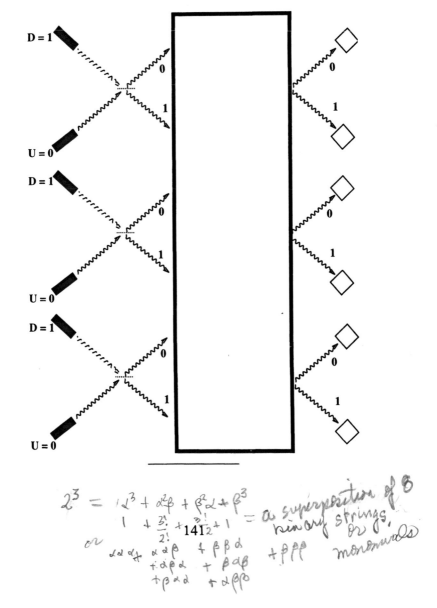

computer to do it in simulation. Likewise, even though our description of the input is an ordinary classical message, what happens next is pure quantum magic.

Suppose the input string is just a string of N zeros (00000 . . . 0). Only photons heading in the U-direction are sent into the array of beam-splitters. After the beam-splitters, each photon is in a superposition of a U-directed state and D-directed state. In terms of our code, the photon state after each beam-splitter is a superposition of a 0 and a 1. It is not simply random, it is qubit and as we have seen that is a different kind of thing. However, every beam splitter is identical, so every U-directed photon is turned into a superposition of a 1 and a 0 state. Now stand back and look at the entire output. What do we have? *The output state is an equal superposition of every possible binary string of length N!* We input a single classical binary string and we get out a superposition of all 2^N binary strings possible! In that first step we glimpse the massive parallelism that qubits open up, for if we now follow the beam-splitters with a qubit manipulator, that is, a quantum computer, we can operate *simultaneously* on all possible binary input strings of length N! In other words, we can run a single computation on all possible inputs in parallel. This massive parallelism was first described by David Deutsch in his seminal paper in 1985.[4]

Quantum gates

Quantum parallelism is not much use if we cannot do anything more than encode the input. To build a computer we need to design physical devices that manipulate qubits to perform logical operations. These devices must be capable of perfectly reversible operation, that is to say, they must be unitary devices. We need to figure out just what to put into the black box of Figure 5.7. I will start with the simplest gate of all, the NOT gate.

I have in fact already introduced you to a unitary NOT gate. The dual arrangement of beam-splitters depicted in

Figure 5.6 is a physical implementation of a NOT gate for photons. In the figure, we code the two mutually exclusive directions of a photon, U or D, as a binary number 1 or 0 respectively. A photon injected in the D-direction emerges with certainty in the U-direction, and conversely, the device performs the logical NOT operation. Two beam-splitters arranged in parallel can perform a NOT operation on photons encoding bits through two mutually exclusive directions. A single beam-splitter may then be thought of as a SQUARE-ROOT-OF-NOT gate, as first pointed out by David Deutsch. Such a gate only makes sense for qubits. It has no classical analogue. A SQUARE-ROOT-OF-NOT gate is the most fundamental operation we can perform on a single qubit. It produces an equal superposition of two logical states. These days it is more common to refer to a SQUARE-ROOT-OF-NOT gate as a Hadamard gate, or H-gate for short, which is at least easier to say. The H-gate is the fundamental single qubit gate. How it is implemented physically will depend on the particular physical encoding of the logical states 0 and 1. We don't have to use two mutually exclusive directions coupled by a beam-splitter. We could, for example, use the two mutually exclusive directions for magnetic orientation that formed the basis of my discussion of entanglement in Chapter 2. It is useful to have a simple way of representing a H-gate which does not refer to the particular physical implementation we are using. The business of quantum computing is not very old and conventions are not well established. However, I think there is some consensus on how to depict qubit operations in terms of *quantum circuits*. This is a kind of picture in which single qubits are represented as lines or 'wires' and gates are boxes that connect one or more input wires to an equal number of output wires.

An example, based on a single beam-splitter, is shown in Figure 5.8. A beam-splitter operates on a single qubit, which we represent as a single horizontal line directed from left to right, as input qubits are transformed to output qubits. It is not necessary for the physical systems representing the qubits to actually move, as in the case of photons. It may be that the physical qubits just stay where they are, and different gates act upon them as time proceeds. We can still represent the transformation of fixed qubits in terms of a quantum circuit. A beam-splitter physically implements a Hadamard gate (H-gate). This is represented in the quantum circuit by a small box, containing a H, somewhere along the wire representing the qubit.

A NOT gate is then represented by a quantum circuit containing two H-gates in series. In Figure 5.6, this is physically implemented by two beam-splitters in series. The quantum circuit for a NOT gate is shown in Figure 5.9. The gate is indicated by a small circuit with a cross through it. The quantum data preparation depicted in Figure 5.7 may be represented by three quantum wires in parallel, each containing a single H-gate (see Figure 5.10). If the overall device is to be physically reversible, it must have the same number of input lines as output lines.

Hadamard gates and NOT gates are not going to be sufficient to build a general purpose quantum computer. Is there some universal quantum gate which can be used to perform all possible qubit transformations on an arbitrary number of quantum wires? The first and obvious thing to

Figure 5.8 A quantum circuit for a single beam-splitter. The diagram is read as if time moved from left to right, so that the input qubit state is on the left and the output qubit state is on the right. The beam-splitter performs a Hadamard transformation of a qubit encoded in the two mutually exclusive directions coupled by a beam-splitter. This is indicated by the box containing the letter H.

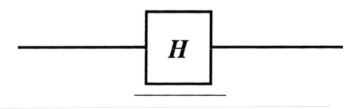

Figure 5.9 The quantum circuit for a NOT gate constructed out of two Hadamard gates in series.

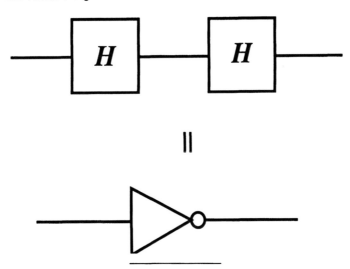

Figure 5.10 The quantum data preparation procedure depicted in Figure 5.7, redrawn as a quantum circuit. Each input line is acted on by a separate Hadamard gate.

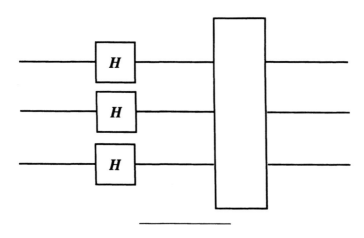

try is to find a physical implementation of a Toffoli or Fredkin gate that operates at the level of qubits. Such a device would need three input lines and three output lines, and as such is a three-qubit gate.

It is not difficult to come up with physical systems, operating by quantum principles, that exactly reproduce a Toffoli or a Fredkin gate. In 1989, I showed how a Fredkin gate could be implemented using photons and beam-splitters, together with optical devices with an intensity dependent refractive index. Independently, Yoshi Yamamoto, then at NTT, came up with a similar proposal at about the same time. Such gates, however, miss the point of quantum computation. Because they exactly mimic classical gates, they miss the opportunity to make use of Feynman's rule to do new kinds of operations. These quantum Fredkin gates do not entangle the states of the qubits upon which they act. What we really want is a universal gate that is capable of all possible qubit operations, including those which lead to a massive entanglement of all the qubits in the system. Such a gate is not just a universal gate for computation, it is a universal quantum gate.

As far as I am aware, it was David Deutsch in 1989 who gave the first suggestion for a universal quantum gate. Deutsch's gate was based on the Toffoli gate and thus acted as a three-qubit gate. It can implement a Toffoli gate and much more besides. If a sufficiently complicated network of Deutsch gates is built, it can realise all possible reversible operations on a set of qubits. In other words, a Deutsch gate can do any manipulation of qubits allowed by Feynman's rule. The dramatic increase in computational power that this leads to may not be obvious to you yet. Be patient, I will shortly give an example.

Since Deutsch proposed his universal three-qubit gate, a succession of physicists have simplified the description. In 1995, David di Vincenzo of IBM Research showed how to construct the Deutsch gate by an arrangement of gates acting on only two qubits. It was subsequently shown by

Tycho Sleator of New York University and Harald Weinfurter from the University of Innsbruck, that two-qubit gates
can be universal for quantum computation. The same result
was proved independently by Adriano Barenco of Oxford
University. This is quite a surprise as we know that for
ordinary classical bits, we need a gate that acts on at least
three inputs. In the quantum case we can get away with
devices that act on any two inputs, provided they represent
qubits. In fact, we now know that almost any two-qubit
gate is universal for quantum computation.

Proving that some qubit gate is a universal quantum gate
is one thing, but connecting such gates together to do
universal quantum computation is quite another. A quantum
Turing machine is a general purpose computer capable of
solving any computational problem involving computable
functions. It would not be much use to give someone a bunch
of universal quantum gates, and then tell them that to
assemble the gates to perform a given computation they must
first perform that computation on some pre-existing universal
quantum Turing machine. As David Deutsch puts it:

> even if a supply of [universal quantum gates] were readily
> available, it has not yet been shown that they would be useful
> to anyone who wanted to perform computations with them.
> For the gates would first have to be assembled into the
> network appropriate for the desired computation . . . and
> nothing I have said so far shows that performing this assembly
> is not tantamount to performing the desired computation.[5]

Fortunately, A.C. Yao proved in 1993 that the Deutsch
type networks of universal quantum gates can simulate a
universal quantum Turing machine to any desired level of
accuracy. Building a universal quantum computer is no
different from building a universal classical computer, so
long as we can find ways to build networks of universal
quantum gates.

There are a number of ways to describe universal
quantum gates. I will use a combination of Hadamard gates,
which operate on a single qubit, together with a particularly

simple two-qubit gate, the *controlled-NOT* (CN) gate. While each of these gates independently is not a universal quantum gate, together they suffice to build a universal quantum gate. The CN gate is actually a classical gate, sometimes called an X-OR or exclusive-OR gate. It has two inputs and two outputs, represented in a quantum circuit by two quantum wires. One of these wires is called the control and the other is called the target. It works like this. If the bit on the control wire is 1, invert the bit on the target wire, otherwise do nothing. In both cases, the control bit is unchanged by the gate. The quantum circuit for a CN gate is shown in Figure 5.11.

A CN gate for photons

A simple beam-splitter can be used to implement a Hadamard gate for photons. Is there a simple way to

Figure 5.11 The quantum circuit for a controlled-NOT gate (CN gate). The top wire is the target and the bottom wire is the control. The direction and nature of the connection of the coupling of the wires is indicated by a vertical line connected to the control wire by a dot and connected to the target wire by a NOT symbol. If the bit on the control line is 1, the bit on the target wire is inverted, otherwise nothing happens. In both cases the bit on the control wire is unaffected.

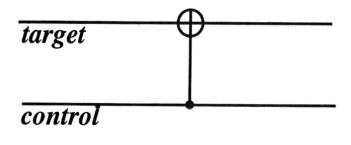

implement a CN gate for photons? Nicholas Cerf, C. Adami of Caltech together with Paul Kwiat of Los Alamos National Laboratory recently suggested how this might be done using polarised light (see Chapter 1). A single photon can be described by an additional physical quantity which can assume only one of two mutually exclusive values in an experiment. This quantity is polarisation. I will denote the two mutually exclusive values as 1 or 0 corresponding to the two mutually perpendicular polarisation directions for light. In Chapter 1, I showed that if the idea of photon polarisation was to be reconciled with the results for experiments with polarised light, we must describe the results of polarisation measurements on a single photon as irreducibly random and described by probability amplitudes not simply a probability. It is remarkably easy to manipulate these probability amplitudes directly. A simple sugar solution can cause the probability amplitudes for polarisation to rotate (see Chapter 1). To make a CN gate for photons we need only a single optical device to rotate the probability amplitude for photon polarisation. Such devices abound and are used in classical optics to rotate the plane of light polarisation.

Nicholas Cerf and his co-workers propose using both the direction of a photon at a beam-splitter and polarisation to encode two qubits of information. We have already seen how a beam-splitter physically implements a qubit in terms of the two mutually exclusive possibilities for transmission and reflection. Let me call this the direction qubit. Now we add the possibility that the photon incident on the beam-splitter can be polarised in two mutually exclusive ways to represent a second qubit. I will call this the polarisation qubit. The polarisation qubit and the direction qubit can be manipulated independently or jointly.

To make a CN gate we use a single beam-splitter to realise the direction qubit. After the beam-splitter, along one of the output directions, we place an optical device to rotate the polarisation qubit by exactly 90 degrees. This means a photon polarised in one direction is converted to

149

a photon polarised in the other, mutually exclusive direction. Suppose we place this device in the path of the reflected (U) photon (see Figure 5.12). The polarisation qubit will only undergo a change if the direction qubit is 0, that is, it is heading in the U-direction. If the direction qubit is 1, that is, the photon is heading in the D-direction after the beam-splitter, the polarisation qubit does not change. We can then regard the polarisation qubit as the target and the direction qubit as the control.

Let's now try and build a simple quantum computer out of direction qubits and polarisation qubits. As an example, suppose we only want to multiply a single input

Figure 5.12 A CN gate for photons as suggested by Cerf et al. The two qubits are implemented by a photon direction and polarisation. For each direction, U or D, the polarisation can be H or V. Only in the case of a photon leaving a beam-splitter in the U direction is the polarisation switched.

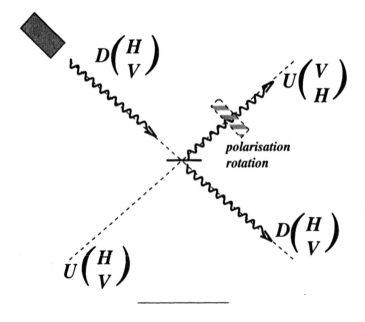

bit by two. In binary code, two times one is not 2, but rather 10. The full range of possibilities is shown in Table 5.3. If we tried to implement this table directly by a device with one input quantum wire and two output quantum wires, it would not be physically reversible, as the *time-reverse* of the device would erase information. To be reversible we need at least as many input wires as output wires. Obviously we need at least one more input wire. This wire will carry a 'preset' bit to configure the gate as multiplication. It is a kind of garbage bit in reverse.

To implement this simple quantum computer, set the preset bit to 0. The logic table must now look like Table 5.4. In fact, the preset bit will contain the carry bit for

Table 5.3 Binary multiplication by two, for one input bits

INPUT	OUTPUT	
0	0	0
1	1	0

Binary multiplication by two table, for one input bits. The output requires two bits as one of the possible results is 2 which has the binary code of 10. As it stands, this table cannot be implemented in a physically reversible way, as the number of inputs does not equal the number of outputs.

Table 5.4 Logic for multiplication of one bit by two

INPUT		OUTPUT	
CARRY BIT	ONES BIT	CARRY BIT	ONES BIT
0	0	0	0
0	(1)	1	0

Same inverts

Target   Control   Target   Control

A logic table for multiplication of a single bit by two. To be reversible we need two inputs and two outputs. One of the inputs is preset to 0. At the output this bit will be the carry bit for 1 plus 1.

151

multiplication of one by two, so in the table it is called the carry bit.

Looking at Table 5.4 it is immediately obvious that a simple CN gate in which the target is the carry bit, followed by a CN gate in which the ones is the target, will suffice. It is just as easy to come up with a classical circuit for ordinary bits to implement multiplication by two. What is so special about doing it with qubits? Well, here is one thing a classical computer cannot do. Suppose we add a single Hadamard gate to the control qubit in Figure 5.11, so that we get the circuit shown in Figure 5.13.

The effect of the Hadamard gate is to prepare the data qubit in an equal superposition of both possible values, 0 and 1. Now the multiplication will act on both inputs simultaneously! In one step we can do two calculations in parallel, simply by preparing the input in a superposition of all possible inputs. As the input is a superposition, the output is also a superposition of both possible results to the calculation. This looks like some kind of parallel computing. We could of course try to do something similar by using two ordinary classical computers, one of which is given the input 1 and the other is given the input 0. However, the qubit multiplication machine can act on both inputs in a single pass through *one device*. Before getting too carried away, however, it is worth noting that when

Figure 5.13 A quantum circuit for performing a controlled NOT using a superposition of both possible control states.

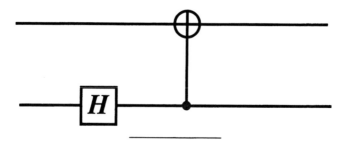

you try and read out the quantum computer you still only get one answer, either 00 or 11. Which answer you get is as random as a coin-toss. What is the use of a machine that produces one of two possible answers at random? The answer is, very useful indeed. To see why let us look at a more interesting problem than multiplication by two, a problem first posed by David Deutsch.

Deutsch's problem

Investment decisions often rest on the outcome of meetings of the Federal bank. For example, suppose the bank is meeting to decide whether or not to raise interest rates. The input to your investment decision rests on a single yes/no result. Whatever the answer, you need to do a long and detailed calculation to make your decision. Suppose, for example, you want to know if the bank's decision has any relevance for the share market. The results of this calculation are also either YES or NO. It does not matter what the actual result of the calculation is for each of the two possible inputs. The only thing that matters is whether or not the results are the same regardless of what the bank decides. What matters is that the result of the calculation for a YES input is *different* from the result of the calculation if the input is NO. It may turn out that the market will rise regardless of the bank's decision, or that the market will fall regardless of the decision. In that case you can start making transactions without waiting for the bank to decide, as the decision is irrelevant. This might be a very useful thing to know, if every other investor in the market is standing around waiting for the bank to decide. While they are paralysed awaiting a decision, you are out there buying or selling to the best advantage.

But there is a problem: the calculation is so long and detailed that when the meeting of the Fed is announced you only have enough time to run the calculation once for a given input. If you can only run the calculation once, you cannot make your decision. To know if the results of the calculation are the same, regardless of what the bank decides, you need to run the calculation once for each outcome to the Fed's meeting, and compare the results. In that case you will just have to wait for the result of the meeting like everyone else.

To put this in mathematical terms, we can represent a YES decision by the bank as a 1 and a NO decision by the bank as a 0. The calculation now computes some function, f, of the input to give one of two results: f(0) or f(1). The results in both cases are also represented as either 1 or 0. It does not matter what f(0) or f(1) actually are. What matters is whether or not they are the same or different. If they are the same, then f(0) and f(1) are either both 0 or both 1. If they are different, than f(0) is the opposite result for f(1). The only way to find out which is which is to run the calculation twice, once for a 1 input and once for a 0 input, and compare the results. However suppose that, for this calculation, even the ultra-computer in New Mexico takes 24 hours to produce an answer to just one calculation, and that is too slow for you to act. But, if you have a quantum ultra-computer, you are in business, as you can prepare the input in an equal super-position of both inputs, 1 and 0, and run the calculation just once.

Here is a quantum algorithm for solving this problem recently proposed by Richard Cleve of the University of Calgary in Canada, and co-workers in Italy and Oxford.

The quantum computer will need two input lines and two output lines. One line is for the input data, labelled x. This line will be changed so as to contain the answer to Deutsch's question at the output, that is, the output is 1 if f(0) is the same as f(1) and 0 otherwise. I will call this line the control-line. The other line will initially carry a preset input, labelled y. This line is unchanged by the operation. I will call this line the target-line (see Figure 5.14). We will need a Hadamard gate to prepare the control-line, x, in an even superposition of 0 and 1. This

154

enables the function f(x) to be evaluated simultaneously for both inputs in a single pass.

At some point we must calculate the function f(x). We must be a bit careful in evaluating this function. If f(0) = f(1) we have a problem, as in that case the calculation is logically irreversible. The two distinct input values, 0 and 1, are taken to the same output value. We can fix this by evaluating a slightly different function, which is called the f-controlled-NOT gate. In this gate the control-line, x, is the control and the target-line, y, is the target. The target bit only becomes inverted when f(x) is 1 rather than when x is 1, as in a controlled-NOT gate. As the controlled-NOT gate is logically reversible so is the f-controlled-NOT gate. The f-controlled-NOT gate must be implemented by a unitary transformation of the target-line. We need not worry about how this is done here. In Figure 5.14 the entire operation is symbolised by a connection between the control-line and the target-line.

The next step is to prepare the target-line in a special state which is physically unchanged by the f-controlled-NOT gate. This would appear to negate the role of the target as something that gets changed by the control. However as we are dealing with probability amplitudes we have more freedom for changing the state of the target than simply flipping its logical status. We can simply change the sign of a probability amplitude. In an experiment, probability amplitudes make their presence felt by determining actual probabilities. The probability is found by squaring the relevant probability amplitude. So if we change the state by changing the probability amplitude, but do not change the square of the amplitude, this means that the probability amplitude for this state can only be multiplied by +1 or −1, as when that is squared to get the probability the sign disappears. Let us assume that if the control is 1, the probability amplitude for the target gets multiplied by $(-1)^{f(1)}$, while if the control is 0 the target amplitude gets multiplied by $(-1)^{f(0)}$.

As the control-line is in a superposition of both possible

155

values, 0 and 1, we do not know which of the two values, f(0) or f(1), is used to control the target-line. But if f(0) is in fact the same as f(1), it doesn't matter that the state of the control-line is uncertain, as the target-line is changed in the same way in both cases. In either case it gets multiplied by 1 or it gets multiplied by −1, and in both cases the target-line and the control-line are physically unchanged by the operation of the gate.

However if f(0) is not the same as f(1), the uncertainty in the control gets carried into the target in a very special way. The state of the target is still physically unchanged, however now the control state *does* get changed as the probability amplitude for 0 gets multiplied by a different sign to the probability amplitude for 1. The result is that the target state goes from being a state which is the sum of two amplitudes (one amplitude for 0 and one amplitude for 1), to a state which is the difference of two amplitudes. So after the f-controlled-NOT gate the target-line itself is unchanged, but the control-line either remains unchanged if f(0) is the same as f(1), or it *is* changed if f(0) is not the same as f(1).

Now put a final Hadamard gate on the control-line. If the state of the control is not changed after the f-controlled-NOT we essentially have two Hadamard gates in series, which is just a NOT gate. Thus the bit on the control is inverted. However if the state of the control is changed in just the way described above, the effect of the final Hadamard gate is to restore the control qubit to its initial value. For example, if the bit on the control is 0, the output bit on the control will be 1 if f(0) is the same as f(1) and 0 otherwise. In one pass we have solved the Deutsch problem. The quantum circuit for the algorithm is shown in Figure 5.14.

The version of the Deutsch problem I have just described suggests the power of using entanglement in a quantum computer, but does not overwhelm us, even though we can get the right answer on a single run of the computation. A much more compelling advantage for this

156

Figure 5.14 The quantum circuit for the two-qubit Deutsch algorithm. The bottom line is the target-line and the top line is the control-line. The control-line controls the target-line by an f-controlled-NOT gate. If the target-line is prepared by passing a one-qubit state through a Hadamard gate, this line is left unchanged by the f-controlled-NOT gate. The first Hadamard gate prepares the input in a superposition of both input values, 1 and 0. The f-controlled-NOT gate then acts simultaneously on both inputs. The final Hadamard gates enable a determination of which type of function was implemented. If the result on the control-line is 1 then f(0) is the same as f(1), otherwise the result on the control is 0 and f(0) is not the same as f(1).

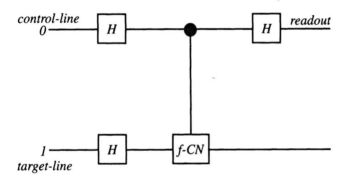

quantum computer arises when we pose the Deutsch problem with more than one input qubit. In this form, it looks a little different.

Ike Chuang and Yoshi Yamamoto of Stanford have a particularly nice way of phrasing the Deutsch problem for many qubits. Alice and Bob live in distant cities and communicate by sending long binary strings to each other, usually referred to as e-mail. To make it simple, let us suppose that Alice's message consists simply of a number from the range 0 to say 999. Bob's job is to read the number and use it to arrive at a yes/no answer to some particular problem. In other words, Bob computes a function, with Alice's number as input and only 0 or 1 as the output. Both Alice and Bob know that this function is one of two kinds. Type 1 functions produce the same output regardless of the input, that is, either all 0 or all 1. Type 2 functions produce an ouput of 1 for exactly half the possible inputs. Alice wants to determine which function Bob is computing, with certainty with a minimum number of transmissions.

In an ordinary classical transmission Alice can only send Bob one number at a time. If Alice chooses her number at random, she will need to send roughly 1000 messages each 10 bits long (as 2^{10} is about equal to 1000). But Alice and Bob subscribe to the new quantum communication service provided by their friendly telecommunications company, so Alice can send arbitrary superpositions of qubits. Instead of sending a single bit string, she sends a uniform superposition of all possible strings, 10 bits long. Bob runs the quantum channel straight into his new quantum computer and pipes the result straight back to Alice, along with the original superposition that she sent. What Alice receives is a superposition of all possible outputs for all possible inputs from one single transmission. In the next step, Alice runs the large entangled state she received from Bob through a very simple quantum circuit to adjust the probability amplitudes in a special way, and returns this result to Bob. Bob now takes this result and runs his previous quantum computation backwards, and sends the new entangled state back to Alice. Alice can now do a simple amplitude manipulation and determine which type of function Bob used, *with certainty*. At the end, Alice has a definite answer, either type 1 or type 2, and has transmitted only two messages, not 1000. If we think of sending a message as a step in the computation, we have reduced the number of steps from about 2^{10} to just two, which is an exponential improvement.

Deutsch's problem was the first example of a problem that could be run more efficiently on a quantum computer.

However, it does seem a little contrived, and was largely ignored by most computer scientists. A few computer scientists and physicists, intrigued by the possibility of massive quantum parallelism, worked out a few more interesting examples where a quantum computer might outperform a classical computer. Still, few people paid much attention. All that changed dramatically one day in 1994, when Peter Shore of AT&T Bell Laboratories at Murray Hill in New Jersey came up with a killer application for a quantum computer: prime factoring.

The financial rewards of number theory

Number theory is an ancient and beautiful branch of mathematics, the playground of mathematical genius for millennia. It seeks to answer questions such as what is the distribution of prime numbers? It seems unlikely that an endeavour of such an abstract kind would be of great interest to bankers, diplomats, or internet merchants. But such is the case, for what all these professions have in common is a need to securely encode sensitive messages, and currently the best method to do that uses the apparent difficulty of finding the prime factors of large composites.

Throughout the ages encryption schemes have been developed to code secret communication. The objective of all such schemes is to ensure that only the intended recipient has the key to decode the message. A scheme is secure if any attempt to crack the code, without the key, is either difficult or impossible. As a child, I often played with codes based on simple encryption schemes that use substitution. For example, a set of 26 numbers can be used to represent each letter of the English alphabet uniquely. Such schemes however are very easy to crack, simply by searching for patterns.

The first unbreakable code was invented by Gilbert S. Vernam of American Telephone and Telegraph Company and Major Joseph O. Mauborgne of the United States Army Signal Corps in 1917. Here is an example of the Vernam cipher.[6] The letters of the alphabet are first coded as one- and two-digit decimal numbers based on a fixed rule known to the sender and receiver. Now break the resulting string of numbers into five-digit blocks, and to each resulting five-digit number add, without carries, another number chosen at random, and transmit only the result of the addition. The set of random five-digit numbers is the key, and must be known by the sender and receiver. Anyone intercepting the message cannot decode the text unless they have the random key. So long as the random key was as long as the message being sent, the code cannot be broken. This means we need to distribute very long random keys every time we need to send a reasonably sized message, a serious practical drawback. The key distribution problem represents the Achilles heel for this kind of code, for if any eaves-dropper gets hold of the key, they can decode the message.

In 1976, Whitfield Diffie, Martin E. Hellman and Ralph C. Merkle, then at Stanford University, discovered the principle of public key cryptography (PKC). Subsequently, in 1977, Ronald L. Rivest, Adi Shamir and Leonard M. Adddelman, then at Massachusetts Institute of Technology, came up with a practical implementation of the idea known as RSA. Public key cryptography overcomes the serious problem of secret key distribution which makes schemes like the Vernam cipher so vulnerable, as in PKC the communicating parties do not need to agree on a secret key beforehand.

To explain RSA, suppose two people, Alice and Bob, wish to communicate in secret, without Eve, the eavesdrop-per, learning what is said. Alice chooses randomly a pair of inverse mathematical operations. One operation can be used for encryption and one can be used for decryption. The particular mathematical operations involved, which for RSA require finding the prime factors of large numbers, are easy to do in one direction, but not the other. Thus, knowing how to do the encryption is not good enough to be able to figure out the instructions for doing the decryp-tion. Alice can broadcast her encryption instructions to everyone including Bob, who can now send a message that only Alice can decode.

The RSA encryption scheme is secure only so long as no-one can find the prime factors of the large number used for encryption. At first sight, that does not sound very secure at all, and indeed for small numbers a systematic search will reveal the factors. Modern high-speed computers could quickly obtain the prime factors of numbers of ten or so digits, simply by a brute search. Suppose the number you are trying to factor is N. The simplest way to proceed is to try and divide N by every number from one up to $\sqrt{N}$. Such a calculation will require, on average, in the order of $\sqrt{N}$ steps of elementary division. If N is a ten-digit number, we will need to perform roughly 100 000 operations, but if N is a fifty-digit number, the number of elementary steps we need to perform is a 1 followed by 25 zeros! The ultra-computer running at 1 teraflop would take in the order of one million years to complete the job on average. That does not sound terribly efficient. The problem is that the number of steps is rising exponentially in the number of digits of the number we are trying to factor. Algorithms like this, that require an exponentially increasing number of steps with respect to the size of the input, are regarded as intractable, for good reason.

However, there is a more efficient algorithm for finding prime factors of large composites. To understand how it works requires a few simple results from number theory. It is not difficult, but number theory is not as widely taught these days as it ought to be, so I will skip the details and just outline the method.[7] Suppose our goal is to find the prime factors of 15. That is an easy problem, and you can see immediately that the answer is 3 and 5. The key step involves something called the *order* of another, randomly chosen number for which 1 is the greatest divisor it has in common with 15. I will call this the seed. Let us choose 11. If we try to divide 11 by 15, we get 0 with 11 remainder. If we divide 11^2 by 15 we get a remainder of 1 (121 divided by 15 goes 8 times with a remainder of

161

1). Let us continue in this way. What remainder do we get when we divide 11^3 by 15? The answer is 11. At the next step, the remainder of 11^4 divided by 15 is back to 1 again, and so it goes. As we raise 11 to successive powers, the remainder after division with 15 alternates between 11 and 1. The *order* of 11, with respect to the remainder of division by 15, is 2. (To see if you have got that, try and show that the order of 7 with respect to the same kind of division by 15 is 4.) Now in the next step, we take 11 and raise it to a power equal to the order and then take the square root. Thus 11^2 is 121, the square root of which is 11. From this number add 1 or subtract 1 to give the pair of numbers 10 and 12. In the final step we find the greatest common factors of 10 and 15 and also 12 and 15. The answers are 5 and 3, and we have found the prime factors of 15. We could equally well have started the same procedure using 7 as the seed instead of 11. In fact for factoring 15, any of the numbers 2, 4, 7, 8, or 13 could be used as a seed. However, if we seed with 14, which does satisfy the condition on the greatest common divisor with 15, the method will fail. Failure of course is easy to check as it is easy to multiply two numbers to see if the result is 15. Because we choose the seed number randomly, there is no guarantee that we would not have chosen 14. The method will thus fail some of the time. However, as failure is easily detected, we just try again with a different seed. For factoring large numbers the method will find the prime factors with a probability very close to one. If we start with a very large number, the seed will also be a large number. The problem with this algorithm is that, for a conventional computer, finding the order of large numbers also requires a large number of steps. Peter Shor's great insight was to realise that finding the order of a number is an easy thing for a quantum computer.

Shor's algorithm for finding the prime factors of a large integer enables an attack to be made on all RSA encryption schemes. That is a very disturbing prospect for a scheme used precisely because everyone thinks finding factors is a difficult problem. A little number theory and a quantum computer, and suddenly all secure encryption schemes are not as secure as we thought. There is no doubt about the number theory part of Shor's scheme, but how does a quantum computer get into the picture? Before we start to lament the demise of RSA encryption we had better check what is required for a quantum computer to factor a large number. We shall see that it requires quantum entanglement on a truly massive scale.

Shor's quantum algorithm

A quantum computer can efficiently find the order of a large number. Return again to the case of factoring 15. We choose 11 as our seed and to find the order of 11 with respect to the remainder after division by 15, we need to step through each successive power of 11 in turn. In the case of 11, we only need a few steps to see that the order is 2. However, if the order had been a six-digit number, we would be in trouble. We could give up with pencil and paper and try and program a classical computer to do the job, but we would need an ever more powerful computer if the order became larger. The problem is that a classical computer is trying not to find the value of a particular function but a property of a function, and that requires evaluating the function many times on different inputs. Finding the order of a function then is seen to be of a similar class to Deutsch's problem, and like that problem is more easily done on a quantum computer.

The input data to the computer are a set of numbers representing the number of times we multiply the seed by itself. We need to do this enough times to ensure that we can see the period emerging. For a classical computer this is just the number of times we run the calculation. With a quantum computer, the key is to prepare the input to the computer as a uniform superposition of all powers up to some suitably large number, which in general is quite large. We also set aside a large number of qubits, preset to 0, that

163

Figure 5.15 A scheme for factorising a large integer N using a random number x. These two numbers are supplied to the computer as initialisation data. The power register contains the numbers a = 0,1 . . . which are used to calculate x^a for a range of values of a. The remainder after division by N is written to a special register which is measured or read out before the DFT is applied. Finally the transformed power register is read out to give the period required.

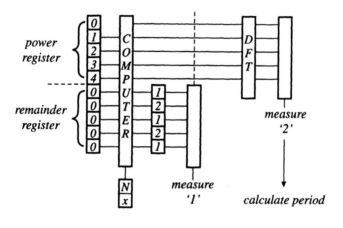

will eventually contain the result of each computation on all the input numbers. I will call these the result qubits. For a moderately sized number that is a lot of qubits. This input is then run through a series of qubit gates to produce an output state which is a highly entangled state of each input number and the corresponding output value for that number in the result qubit. The seed number and the number we are trying to factor are given as preset qubits to the computer. The idea is illustrated in Figure 5.15.

The result of the calculation on each input number is not unique. For example, in the case of a seed of 11 to factor 15, the results of the calculation just oscillate from 11 to 1 over and over again. This means there will be many input states that are entangled with the same output states. If we measure a given output qubit there are many indistinguishable ways this can happen. For example, if we start to factor 15 with a seed of 7, the input qubits will need to encode a superposition of all numbers from 1 to say 20. The output qubits will contain the set of numbers {7, 4, 13, 1} repeated five times. If we measure the output and get the result 4, there are five different input numbers that give this result: 2, 6, 10, 14 and 18. That is to say, 7 raised to each of these powers, when divided by 15, will have a remainder of 4. As the state before the measurement was entangled, the state after the measurement is a superposition of all possible ways the input can produce a 4 in the output register. This must be a sum of five distinct probability amplitudes, corresponding to each of the numbers 2, 6, 10, 14 and 18. Notice that the period we seek, 4, is just the difference between any two consecutive numbers in this superposition. It would not matter what result we got for the measurement. In every case there will be five different ways it could be achieved, each corresponding to a number separated by the period we seek. The net result is to encode the period we are trying to find in terms of a set of superposed probability amplitudes, each corresponding to the different inputs which yield the same number in the result register. This completes the first stage of the Shor algorithm. The next stage is to try and process the probability amplitudes in such a way that a measurement on these qubits will yield the period.

We know that in our example there are five distinct numbers superposed. We do not know what these numbers are, but we do know they are separated by the period we seek. How to get at this period? The first point to note is that the superposition of the five numbers is uniform. This means that if we measure what numbers are in the input qubit register and, given a result of 4 on the output qubit register, we will get each of the five numbers {2, 6, 10, 14 and 18} with equal probability. A single result will not suffice to determine the period. So it is no good simply making this measurement directly. How can we transform these amplitudes so that a readout of the register will yield an integer from which we can infer the period?

To see how this is done recall that before we began we started with the state 0 in the input register and then acted with a set of single qubit Hadamard transforms to produce an input register which was a uniform superposition of all the numbers 0 to 19. We are now going to try and undo this transformation. The undo transformation is rather more complicated as we are no longer dealing with just the 0 state. Technically this transformation is called a discrete Fourier transform, or DFT for short. Instead of starting with the state 0 we now have a uniform superposition of five states corresponding to the numbers {2, 6, 10, 14, 18}. Each one of these will get transformed to a uniform superposition of the twenty possible input states (see Figure 5.16).

After the DFT we can look again to see what numbers are present. Will all possible states occur? For example, we might ask for the probability to find the state corresponding to the number 5. Referring to Figure 5.16 we see that there are five indistinguishable ways to get any final state, as it could have come from any of the five states corresponding to the numbers {2, 6, 10, 14, 18}. Now comes the essential point. To work out the probability of getting to, say, state 3, after the DFT we need to add the five probability amplitudes corresponding to the five ways we can get to state 3. Now, it is possible that these amplitudes could cancel in part or even in total. The DFT is carefully constructed so that for most of the cases, these amplitudes all cancel exactly. In fact, this is what happens for state 3, which means that after the DFT we will *never* find the input register in state 3. In the other cases, however, the five amplitudes all point in the same direction, leading to a high probability. The state corresponding to 0 is a case in point. The other states for which cancellation does not occur are 5 and 15. So after the DFT transformation we are equally likely to find one of the states corresponding to the three numbers {0, 5, 15} and no others.

166

Figure 5.16 The five states {2, 6, 10, 14, 18} remaining in the output register after a measurement of the result qubits, are mapped onto the original twenty states by the DFT. There are thus five different, but indistinguishable, ways to get to any of the final states. In most cases the probability amplitudes for these different ways cancel. In three cases, 0, 5 and 15, however, the probability amplitudes all add up to ensure the only results we will obtain when we finally measure the output register after the DFT are 0, 5 and 15.

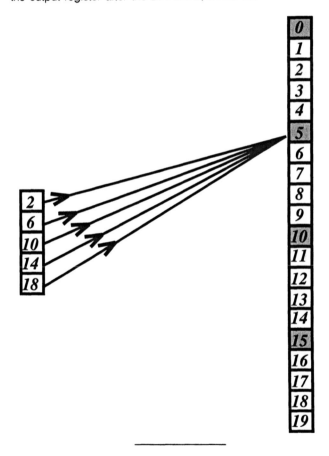

The rule is this: the only states for which the five probability amplitudes do not cancel are those which are a multiple of the number of states (in this example 20) divided by the separation of the five numbers {2, 6, 10, 14, 18} that is, the period. So, the only states that get through are those which are multiples of 20 divided by 4, that is, multiples of 5, including the 0 multiple. When we measure the input register after the DFT we know for sure we will get a number which is just an integer multiple of 20 divided by the period. It is then a trivial matter to work out the period. The algorithm will work regardless of how many states we start with, which of course we know, and whatever the period happens to be.[8]

You may have found the foregoing description of the Shor algorithm a bit hard to follow, so I think I had better summarise the basic steps. First the mathematics. We start with a number N, that we are trying to factorise. We do this by first choosing a large number, y, at random, but making sure that y and N have only 1 as the highest common divisor. Now we want to raise y to successively higher powers and divide by N and keep the remainder. Given enough repetitions of this calculation we can work out the period, r, of this calculation. Once we know the period we can quickly find the factors of N.

Now for the computation. We want to calculate y, y^2, y^3, and so on, divide each by N and keep the remainder. The input numbers are thus the successive powers 1, 2, 3, 4 . . . In step one, we use Hadamard transformations to produce a register of input qubits in a uniform superposition of all these powers. The output qubit register is preset to zero. In step two, we 'run the computer' to write the remainder, corresponding to each power, into the output register. This only takes one run as all possible powers are superposed in the input states, so all the calculations are done in parallel. In step three, we read out the output register. It doesn't matter what result we get for this. All that matters is that the input numbers that correspond to this result will be separated by the period we seek. The input register is then in a uniform superposition of just those powers. In step four, we apply a DFT transformation to this special superposition of input numbers and then readout the input register. The only numbers we will obtain will be those which are equal to an integer multiple of the total number of basis states, divided by the period. Once we have one of these we can work out the period, as we know (by how we built the quantum computer) how many input number states we used.

The key feature of quantum theory that makes the Shor algorithm work is the ability to superpose a very large number of states, and act on all of them at once. This is, of course, just Feynman's rule in action. We will need a very large number of states to code the input and output for a moderately large factoring problem. In fact, the number of states we will need will increase exponentially as the size of the number we are trying to factor is increased. This simply reflects the fact that in a classical computer we would need an exponentially increasing number of steps. In the quantum case all these steps are done at once on an exponentially increasing number of input states. Fortunately the number of available states also increases exponentially in the number of qubits we can use. In fact for k qubits, the number of states we can get to is 2^k. Where a classical computer uses exponentially many steps, the quantum computer uses a superposition of exponentially many states. That is Feynman's rule on a massive scale. That is what gives quantum computation its power.

Shor's algorithm vindicated the belief of Feynman and Deutsch that a quantum computer could solve problems more efficiently than a classical computer. Since Shor took his algorithm to the world, there has been a lot of work on finding other interesting problems that might be more efficient on a quantum computer. As of this time, there is only one other interesting example discovered by Lok Grover of AT&T Bell Laboratories at Murray Hill in New Jersey. Grover's algorithm is a 'quantum search for a needle in a haystack'. The problem is this. Suppose you seek a phone number in a directory of N names. Easy? Yes, for the usual kind of phone directory. Suppose, however, the names in this directory were arranged in a random order. It is not hard to see that to find a particular person's name with a probability of 50 per cent you would need to read at least half the entries. In other words, you would need to access the database at least N/2 times. If, however, we had a quantum database in which the registers were in a uniform superposition of all possible entries, we could extract a particular entry with a probability of 50 per cent by accessing the database only in the order of $\sqrt{N}$ times. This is not quite the exponential speed-up of the Shor algorithm, but it is impressive enough. The search is now on for the class of problems that can be efficiently solved by a quantum computer. We are still a long way from knowing just what kinds of problems belong to this class, but the Shor algorithm and the Grover algorithm give us confidence that it contains at least a few interesting problems. These algorithms have given us all the reasons we need to try and build a quantum computer. It is time to turn from quantum software to quantum hardware.

CHAPTER SIX

THE DREAM MACHINE

At the time of writing, no quantum computer exists anywhere in the world. The power of quantum parallelism, as manifest in the Shor algorithm, remains a dream waiting to be realised. But there are a number of laboratories in hot pursuit of the dream. In laboratories around the world, graduate students labour long into the night in a quest to build a prototype device that would demonstrate the power of quantum computing. There is no shortage of ideas when it comes to how to build a quantum computer. The problem is not lack of imagination. Rather the difficulty of building a quantum computer is due entirely to the requirement that the device operate in a completely reversible way. Unfortunately the world does not run backwards quite as easily as it runs forwards.

Turning time's arrow

It takes little more than casual observation to come to the conclusion that there is a certain direction in physical processes. A hot cup of coffee placed in a room spontaneously cools. A fire slowly dies, leaving behind a cold residue of ash. Data written to a floppy disk, stored in a box, becomes corrupted with errors. People age and die. The sun, like all stars, burns its nuclear fuel to exhaustion, eventually ending its days as a cold lifeless lump of matter. The entire universe itself appears to be slowly and inexorably marching to a cold, uniform distribution of matter. As far as we know, none of these processes is ever going to spontaneously reverse. Time has an arrow, and its flight is irreversible. Time's arrow is aimed directly at quantum entanglement, the heart of a quantum computer.

Despite this fact we believe that the laws of nature are themselves ambivalent as to the direction of time. Quantum physics is no exception. The reversibility of quantum motion is captured in the idea of a unitary transformation, and is essential to the possibility of quantum computation. To build a quantum computer we will need to run against the tide of the universe. As you might expect, that is no easy matter. It is difficult to arrange even a classical system to run in an entirely reversible manner if the system is composed of a large number of particles. It is far more difficult to achieve the same with a large number of quantum systems, such as an array of qubits. This is a serious obstacle to building a quantum computer and solving it takes us straight to one of the most debated aspects of quantum physics, the measurement problem.

Of course classical computers are not protected from the arrow of time. Errors creep in along the way as the computer manipulates bits of information. Parts of the world external to the computer get mixed into the computation and, in effect, rewrite bits. Errors erase information and are thus due to physically irreversible processes. But as we have seen, all current computers are inherently irreversible anyway. Digital computers only work at all because powerful error correction techniques can be used to fix errors. The fact that digital computers only work with two values, 1 and 0, makes them particularly robust. If a voltage on a transistor strays from the correct voltage coding a 1, it does not matter too much so long as the voltage change is not so large as to turn a 1 into a 0. Digital devices are inherently stable. However, occasionally errors do occur and they need to be detected and corrected. In an ordinary digital computer this is not a problem. Resetting a bit, whether it is an error or not, necessarily erases information, so error

172

correction in a classical computer is likely to be a physically irreversible operation. But as a classical computer is already physically irreversible that is not a problem. In effect, friction stabilises a classical computer. We do not have this option for a quantum computer. Any bit error cannot be corrected by an erasure step, as that will certainly entail the Landauer energy price.

If a quantum computer is to run in a reversible way it must remain isolated from the outside world during the computation. At no stage can the computer leave a record of which computational path it has followed, a record that a passing classical observer may or may not read. At no stage can the interaction of the quantum computer with the outside world enable a 'which-path' determination to be made. This is known as *the decoherence problem,* and it requires a very special kind of isolation of the qubits used in a quantum computation. We can get an idea of why the reversible dynamics of a qubit is so difficult to achieve by considering again the simple controlled NOT (CN) gate. Suppose that the control line is a qubit that forms part of a much larger computation and that the target is some other qubit, an alien qubit, that takes no part in the computation at all (see Figure 6.1).

In a CN gate, of course, the bit on the control line never changes, so you might think that having our internal qubit interact with an alien qubit in this way is not a problem at all. For a classical computer it would not be a problem, as the bit on the control of a CN gate is never changed. But for a qubit, we run into a problem whenever the internal qubit is in a superposition of the two logical states. Suppose the alien qubit is set to 0. As this qubit is the target for the internal qubit of interest, it will be changed whenever the internal (control) qubit is in logical state 1. Suppose the internal qubit is in an equal superposition of 1 and 0. This means we know only that a 1 and a 0 are equally likely, but we have no way to distinguish them. Of course if we measure the logical state of the internal qubit we get a 1 or a 0 with equal probability. But if we do not

173

Figure 6.1 A representation of the decoherence problem pictured in terms of a CN gate. One qubit inside a quantum computer leaves a footprint of its logical state in the world outside the computer. If the internal qubit is in a superposition state, the entanglement with the alien qubit outside destroys the superposition by removing the conditions for Feynman's rule. The two superposed states of the internal qubit can now be distinguished by monitoring the alien qubit. This will happen even though for a controlled NOT gate, the internal qubit is the control bit which classically is unchanged by the gate.

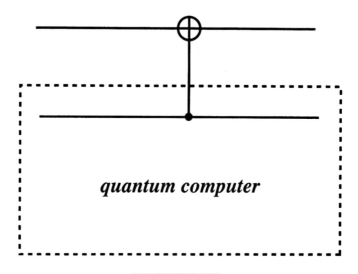

quantum computer

measure it, and continue to process it in a quantum computer, the superposition state will remain. But when this superposition qubit is a control bit, connected to another alien qubit in a CN gate, the target qubit will obtain information on which logical state is present, for if our qubit is a 1, it will change the state of the target qubit. We only need to look to see if the target qubit has changed, and we have determined which of the previously indistinguishable possibilities in the internal qubit were realised.

Why is this a problem? Because the qubit inside the

computer is turned into a coin-toss for all future observations. It is no longer described by Feynman's rule, as the two states are distinguishable; we need only look at the alien qubit outside the computer. It does not matter whether we look at the alien qubit or not. The mere fact that we can look at it and decide which of two previously indistinguishable alternatives is realised destroys the superposition state. This must be so, as the future behaviour of the quantum computer must be consistent with whatever we see in the alien qubit, no matter when we look at it. Turning a qubit into a coin-toss inside the computer is going to play havoc with subsequent quantum manipulations. The destruction of superposition states will propagate from the affected qubit to every other qubit it comes in contact with. Our quantum computer will be turned into an enormous (and enormously expensive) roulette wheel. We learn from this example that it is possible to couple our computer to the outside world in a way which for a classical computer makes no difference at all, but for a quantum computer is totally disruptive. The lesson is this. It is not enough to ensure that bits are unchanged by the outside world. We must ensure that no bit leaves a footprint, no matter how faint, in the outside world. Isolating a quantum computer from the outside world is even more difficult than isolating a classical computer from errors.

Quantum error correction

The year 1996 was a watershed year for quantum computation. The year began with the first hint that quantum error correction would be possible. Despite this, considerable pessimism remained. By the year's end, at a conference in the Institute for Theoretical Physics at the University of California, Santa Barbara, optimism reigned. Most participants were convinced that errors no longer presented an insurmountable barrier to building a quantum computer. The rapid progress was tracked by the relatively recent phenomenon of the e-print archive at Los Alamos National

Laboratory (LANL). This archive is the handmaiden of the Internet and has changed forever the publication of scientific results. When an author completes a piece of work, a paper reporting the results is uploaded to a server at LANL. It is then indexed and stored in an appropriate directory, such as quant-phys, for quantum physics results. New papers are posted daily or you can arrange to be notified of new postings by e-mail. Within a day or two of posting, your results are available worldwide, to be downloaded by anyone interested in the subject. For the communities that use it, the LANL archive has changed work methods. Most authors still submit their papers to a refereed journal for eventual hardcopy publication, however, some journals now simply ask that you submit the address of your paper at the LANL archive. Throughout the year 1996 the electronic publication archive at Los Alamos National Laboratory ran hot with new ideas for dealing with errors in a quantum computer. At one stage I was becoming more than a little overwhelmed by the volume of new results arriving in my e-mail account each week. Every Friday afternoon I would download papers of interest which then consumed my entire weekend. It was a busy year, which ended on a rather more relaxed note during the Institute of Theoretical Physics program on quantum computation at University of California, Santa Barbara. By the end of 1996, powerful new principles had been discovered that enabled quantum computers to be protected from errors—both classical bit-flip errors, and superposition destroying errors.

To explain how these methods work would require a long digression on classical error correcting codes for binary strings. The key idea, however, is redundancy, that is, to use strings of bits to encode a single logical 1 or 0. These bit strings are called codewords. Such codes are a well developed part of classical computer design. For a quantum computer, we need a new variation on the theme if we are to protect superposition states. The new idea is to use *superpositions* of codewords to encode a single logical 1 or a 0. In that case a superposition of a 1 and a 0 can be a very complicated superposition of many states corresponding to different codewords. In the words of John Preskill at Caltech, we can use entanglement to fight the entanglement with the outside world caused by errors. Quantum error correcting codes were discovered independently by Peter Shor of AT&T Bell Laboratories in New Jersey, and Andrew Steane of Oxford University, in late 1995. Within a few months it was shown that such codes could correct a great variety of errors. The discovery of quantum error correction saved quantum computing from the trash can.

In fact, it is not necessary to get a 100 per cent recovery from errors in a qubit device. Seth Lloyd of Massachusetts Institute of Technology and Charles Bennett and co-workers at IBM Research showed in 1996 that, under the right conditions, we can recover a quantum state from an error with high fidelity if the error per qubit is less that about 19 per cent. Unfortunately 'the right conditions' requires that we encode the information, detect and recover from the error without making a mistake. That is rather unrealistic, as encoding and error detection are themselves complex qubit manipulations that are quite prone to error. If quantum computation is to be robust we must be able to recover from errors both in the computation and in the error correction process itself. In other words, we must be able to do *fault-tolerant* recovery. In May of 1996, Peter Shor showed that we could achieve that aim. Provided the error rate is not too high, fault-tolerant quantum computation is possible in principle. Subsequently in the Caltech group of John Preskill, Dan Gottesman combined the ideas of Shor with recent work of Knill and Laflamme at LANL to show that there is a threshold probability for error above which quantum computation can be fault-tolerant. The new schemes for fault-tolerant quantum computation can handle an error rate of the order of one in one million, provided that an error occurs one bit at a time. This is big enough to encourage experimentalists to try and build a quantum computer using available technology. The down side, however, is that fault-tolerant computation requires many more qubits than is strictly necessary to do a particular quantum computation, such as factoring. Fortunately the number of qubits required appears not to scale exponentially with the minimum number of qubits required to do a given computation, so there is no fundamental obstacle in using fault-tolerant computation. However, trying to build networks for such a massive qubit entanglement is a very grand challenge for any laboratory scheme.

There are a number of design proposals based on different physical realisations. I will describe the three that I think have the most chance of success in the near term. Of course, any day some wonderful new proposal may arrive on quant-phys at the LANL archive, so in the end none of the current schemes may be used. However, it is worth looking at these three proposals simply to appreciate the enormous challenge that building a quantum computer poses.

The stuff of dreams

Have you ever seen an atom? In the laboratory of Dave Wineland at the University of Colorado in Boulder you can watch a single ion blinking uncertainly in the glare of a laser beam. This represents the culmination of a 30-year effort, in many laboratories around the world, to trap ions in strong electric fields. In the Boulder trap a single ion of beryllium is held in the tight embrace of oscillating electric fields on confining electrodes (see Figure 6.2).

By a combination of laser light and electric fields, the ion can be slowed down so that its position and velocity are as tightly constrained as is possible in the face of irreducible quantum uncertainty. In this state, its motion can be choreographed by sequences of laser pulses, without ever leaving the quantum domain.

Within the ion, the electrons also obey the quantum rules, which ordain that electrons must only exist in a discrete set of states each with a particular energy. Two of these states are special as they correspond to electrons that can be pushed around by exciting laser pulses. Each of these states has a

Figure 6.2 A schematic representation of the arrangement of electrodes used to trap ions in the laboratory of Dave Wineland in Boulder. This arrangement is called a linear Paul trap. The ions behave as if constrained by springs in all three spatial dimensions.

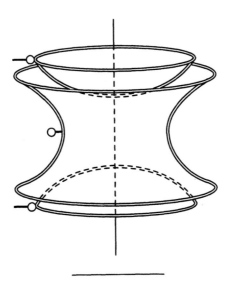

definite energy of motion, and the exciting laser pulses can move an electron from one of these special states to another, or produce an equal superposition of both states as required in a Hadamard gate. The laser excitation causes the atom to tremble a little within its electric prison, a delicate nudge, to set it vibrating in the trap. In this way, laser pulses can cause the electronic states to become entangled with the vibrational motion of the entire atom. The two mutually exclusive electronic states of each ion are used as a qubit. So N ions can be used to realise N qubits. The trick, however, is to couple together at least two qubits at a time to realise elementary universal quantum gates.

In 1994, Ignacio Cirac, then at Universidad de Castilla-La Mancha, and Peter Zoller, of the University of Innsbruck in Austria, realised that if more than a single ion were held in the trap, it would be possible to entangle the motion of all the ions and the electronic states of each individual ion. On the basis of this idea they went on to show how to do elementary quantum qubit manipulations such as Hadamard and CN gate transformations.

Cirac and Zoller proposed that a number of ions be confined to a line in a suitable arrangement of electrodes. The ions are held in the trap by the external electric fields imposed by the electrodes, and line up so as to minimise the electric repulsion between them. The arrangement resembles a set of railway cars connected by rigid rods. The entire assembly of ions can be cooled so that only one collective vibrational state is important. This vibrational state corresponds to all the ions moving as one, backwards and forwards in the line. It is possible to shine laser light on the entire assembly and image each individual ion. The resulting image looks exactly as you would expect, a set of little lights lined up along the axis of the trap, blinking on and off as photons from the illuminating lasers are scattered towards the camera. If we can see individual ions we can separately target each ion with an appropriate laser beam (see Figure 6.3).

When one ion interacts with a laser pulse it can set the entire chain in motion. However, as the entire chain is moving according to the rules of the quantum, we now find that it is possible to entangle the collective motion of all the ions with the electronic state of any one of them. In effect, the vibrational motion acts like a 'quantum bus' to ship quantum information from one electronic state to another, carefully controlled by targeting ions with laser pulses. In 1995, Chris Monroe, Dawn Meekhof, Brian King, Wayne Itano and Dave Wineland at the Boulder National Institute of Standards and Technology (NIST) laboratory managed to implement a CN gate with only a single trapped ion.

Each quantum gate in some circuit for a complex computation may involve a sequence of one or more laser pulses targeting a particular sequence of ions. The DFT

Figure 6.3 A linear ion trap. The four long cylinders represent electrodes to keep the ions in a line and end caps stop the mutual repulsion of each ion from blowing the arrangement out the ends. A set of three laser beams is depicted selectively exciting particular ions in the trap.

part of the Shor algorithm with a total of K qubits would require $K(2K-1)$ laser pulses. The real bottleneck in the Shor algorithm is the part of the algorithm which computes the remainders of successive powers of a number. David Beckman, Amalavoyal N. Chari, Srikishna Devabhaktuni and John Preskill of Caltech have shown that to do this part of the algorithm would require about $396K^3$ laser pulses. Quantum error correction can itself require many laser pulses. Sam Braunstein of the University of Wales at Bangor has shown that a five-qubit code, implemented in an ion trap quantum computer, would require about 24 laser pulses.

The ion trap quantum computer is an amazing machine which can produce an entangled state of the vibrational motion of many ions. However, it suffers from a number

181

of serious problems. The first of these is the 'clock-speed problem', in other words, it is very slow to run. Even under the best of circumstances it is unlikely to be possible to perform more than about 10 000 gate operations every second. Given the very large number of laser pulses required, this is going to make any reasonably interesting quantum computation rather slow and probably very much slower than the rate at which errors will occur. The second major problem is connected with the first. To speed up the run time it would be a good idea to use shorter and shorter laser pulses. However, to have the same effect a short pulse will need to be more intense. Unfortunately, if the laser pulses are intense, the computer will suffer what John Preskill has called 'qubit leakage'. The problem is that for intense laser pulses more than two electronic states may be involved during the excitation. The electronic qubit system is no longer a single qubit. This problem was first identified by Martin Plenio and Peter Knight of Imperial College, London. Unfortunately, this appears to be an almost insurmountable problem for any practical ion trap quantum computer. Plenio and Knight suggest that an ion trap quantum computer will be hard pressed to factor 15—a rather modest achievement to say the least. Last but not least is the daunting experimental challenge of trapping and cooling a large number of ions in a linear trap and still have the ions following a simple collective motion. More likely the ions will rearrange themselves into a kind of braided string, the motion of which is likely to be rather difficult to control.

Despite its possible limitation as a quantum factoring device, the ion trap quantum computer may be useful in other areas. An entanglement of only four ions would still be a considerable achievement, as to date quantum features of entangled systems have been tested for only two qubits (the tests of the famous Bell inequalities). Teleportation and GHZ experiments would need at least three-qubit entanglement and have only just been realised by the LANL group (see page 84).

182

Clock-speed is a serious problem for current ion trap quantum computer schemes. Another proposal which can do much better in this respect is from a field known as *cavity QED*. The cavity in this name refers to an optical cavity, which is simply a bucket for photons. It is easy to make a leaky photon bucket. Take two concave mirrors and set them up opposite each other (see Figure 6.4). A photon is bounced back and forth between the mirrors. However, as no mirror is perfectly reflecting eventually the photon will escape. It will do this at a random time, but the better the mirrors, the longer is the average time to escape. The QED stands for quantum electro-dynamics, which is the field of physics dealing with the quantum description of light interacting with atoms. Inside the cavity we place an atom, which can periodically capture and release the photon in the cavity. If the cavity is not very leaky the photon can be captured and released many times. The trick, of course, is to ensure that when the atom releases the photon, it emits it back into the cavity and not in some arbitrary direction which can cause it to escape the cavity altogether. This is the *spontaneous emission problem*. If that happens, effectively the photon bucket has developed another hole.

A huge experimental effort over the last twenty years has enabled cavity QED to work at the level of single atoms and single photons. Progress has been achieved by making very good cavities that hold a single photon for a long time and using atoms that nearly always emit the photon into the cavity. A world leader in this field is the group of Serge Haroche at Ecole Normal Superiáure in Paris. The mirrors in the Haroche experiments are made from highly polished pieces of superconducting niobium. The photons are in the microwave region of the spectrum, so this is a microwave cavity. You probably have a microwave cavity at home used for cooking. Needless to say, microwave oven cavities use a lot more than one photon. It would take a very long time to heat a frozen pizza one photon at a time. It is not easy to hold an atom inside an optical cavity. In the Paris

Figure 6.4 A cavity QED experiment consists of two almost perfect mirrors placed very close together. Atoms are either injected through the cavity or are trapped inside. An atom can dump a photon into the cavity, and reabsorb and re-emit it a number of times, before the photon eventually leaks out. Alternatively, the atom may simply change the refractive index inside the cavity, thus changing the effective length of the cavity as seen by any light present inside. Additional laser beams can be provided to pre-excite the electronic state of atoms entering the cavity, or to post-process the electronic state before detection. Finally, a special detector can determine if the electron in the atom is in the ground state or the excited state, the final readout of two mutually exclusive possibilities required for all qubits.

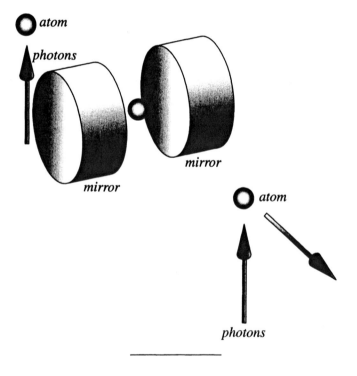

experiment, atoms are injected into the cavity from the side. While they fly through the cavity, however, they can still capture and release a photon many times, so this is almost as good as keeping the atom fixed inside the cavity. As in the ion trap experiments, only two electronic energy states participate in the exchange of a photon with the cavity. These two levels are used to encode the qubit. A similar experiment is underway in the laboratory of Jeff Kimble at the California Institute of Technology (Caltech) in Pasadena. Kimble, however, uses light of a much higher frequency and his cavities are very tiny. The mirrors are only $\frac{1}{100}$ of a millimetre apart. This helps achieve many exchanges of the photon between the atom and the cavity. Like Haroche, Kimble also uses two special electronic states in the atom to encode the qubit. In their early experiments, the Caltech group simply dropped cold atoms through the cavity, so the qubits physically moved through the device. Kimble called these *flying qubits*. The Caltech group is now constructing an experiment in which a single atom will become trapped inside the cavity. It is held in place by the weak force exerted on the atom by the atoms in the mirrors themselves, or simply by the very weak light beams due to a few photons in the cavity.

Both the Paris and Caltech cavity QED experiments manipulate qubits in a manner very similar to the ion trap. Instead of the electronic state of the atoms being entangled with a collective vibrational mode, they become entangled with a single photon inside the cavity. The state of the flying qubits can be manipulated by other laser pulses before they enter the cavity and can be post-processed by laser pulses after they leave the cavity. It is also essential to carefully control the velocity of the atoms as this determines the time during which the cavity and atom can exchange a photon. A scheme for implementing a CN gate is shown in Figure 6.5.

It works by sending three successive atoms though the cavity. The first atom encodes the control qubit. It is prepared in a superposition of the two electronic states. If

185

Figure 6.5 A scheme for implementing a CN gate for flying qubits in a cavity QED experiment, as proposed by the group of Serge Haroche in Paris. Atom 1 is the control bit. If it is in the ground state it does nothing to the cavity, but if it is in the excited state it leaves a photon in the cavity. If it is in a superposition of both states the cavity is left in a superposition of 0 and 1 photon. The target qubit is atom 2. It interacts with the field via a different mechanism, so that another laser pulse, S, will only switch the state if there is a photon in the cavity. Finally a third atom resets the cavity.

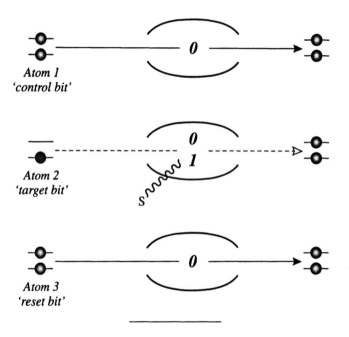

Atom 1
'control bit'

Atom 2
'target bit'

Atom 3
'reset bit'

the atom is in the ground state, it does nothing to the cavity. If the atom is in the excited state it deposits a photon in the cavity and returns to the ground state. The effect is to put the cavity in a superposition state of no photon and one photon. A second atom carries the target qubit. The state of this atom is changed conditionally on whether or not the first atom placed a photon in the cavity.

The change is made by another laser beam, the auxiliary, directed through the side of the cavity. If there is no photon in the cavity, the auxiliary laser has no effect on an atom passing through the cavity. If there is one photon in the cavity, it shifts the allowed energies of the atom which enables the auxiliary field to invert its atomic state. Finally, a third atom clears the cavity state, ready for the next operation.

The advantage of cavity QED schemes is that the time scale for the interaction between qubits is much shorter than for ion traps. Thus more gate operations can be done per second. The down side, of course, is the need to physically move the qubits through the cavity in a highly controlled manner. The Caltech experiment may solve this problem by trapping the atoms inside the cavity. For a long sequence of gate operations a very large number of flying qubits will need to transit the cavity, placing huge requirements on the stability of the operation.

Finally, I will describe the most unusual proposal of all, the quantum computer in a test-tube. Actually, I should say quantum computers in a test-tube, for this scheme is based on using each individual organic molecule in solution as a quantum computer. The idea, due to Neil Gershenfeld of MIT and Ike Chuang of Stanford University, makes use of the mature technology of nuclear magnetic resonance (NMR). Earlier work by Seth Lloyd, also of MIT, had suggested the possibility of using the magnetic orientation of the nuclei within individual molecules as qubits. What Gershenfeld and Chuang realised, however, is that it is not necessary to isolate an individual molecule; we can work with an Avogadro's number of molecules, that is, 10^{23} quantum computers in a test-tube!

The idea is to use the magnetic orientation of certain nuclei in complex molecules. As an example, consider a two-qubit system made from (2,3)-dibromothiophene (see Figure 6.6).

The two hydrogen nuclei are simply individual protons. The magnetic orientation of a proton in the molecule is

Figure 6.6 The molecule (2,3)-dibromothiophene contains two hydrogen nuclei, protons, that each have two mutually exclusive magnetic orientations (spin). The orientation of each proton in an external magnetic field can thus be used to encode one qubit. External radio frequency fields can be used to produce superposition states of the qubit to realise a Hadamard transformation. The nuclei also influence each other through their individual magnetic fields. This influence can be exploited to make a two-qubit gate such as a CN gate.

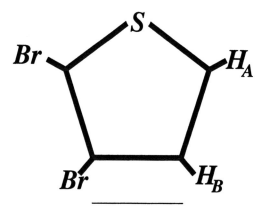

restricted to just two values: up or down with respect to some external magnetic field. This kind of orientation is called spin and formed the basis of my discussion of entangled states in Chapter 2. We thus have an ideal qubit in each of the two hydrogen atoms, allowing us a two-qubit system. In large, complex molecules there may be hundreds of nuclei that can be used in this way. Even as few as ten nuclei would enable us to prepare superpositions of 2^{10} binary strings, that is, about 1000 distinct states. Furthermore, it was noted long ago by the founding father of NMR, Felix Bloch, that nuclear spins are particularly immune from the problem of decoherence. To Bloch this appeared to be a problem, as it made extracting a signal difficult. For a quantum computer designer it is a feature to be exploited. Nuclear spins ignore time's arrow—at least for long enough to get the computation over and done with.

It sounds too good to be true. The ion trap guys are labouring away trying to make a single CN gate and here is a proposal to make, not one quantum computer, but millions upon millions all at once. Indeed, there is a big problem with using a test-tube full of molecular quantum computers. The spins on different molecules are all over the place. While each molecule is a quantum computer, each quantum computer is in a slightly different state. If we could somehow cool all the molecules down, we might be able to get more of them to look the same, but that is a very difficult thing to do as the ion trap experiments have shown. At room temperature the random alignment of spins from molecule to molecule means almost as many spins are pointing down as are pointing up. Almost, but not quite, and that is the point. There is a small deviation from a completely randomised orientation over all the molecules. Gershenfeld and Chuang show how this small deviation from equilibrium can be exploited to do quantum computation. They have discovered a way to actually use the massive redundancy inherent in having a solution of 10^{23} independent N-qubit quantum computers. The principle behind the idea is that there may be considerable structure in the nuclear spins in a complex molecule. Their method relies on the enormous advances in NMR technology over the last 40 years. Exquisitely timed pulses of varying electromagnetic fields can be used to process nuclear spins on molecules. Using such techniques, time's arrow can be halted, at least for some of the sample for some of the time.

But this redundancy comes at a price. Selecting out a small preferred set of molecular spins from a background of total randomness, means the NMR signal from that small set will be tiny compared with a background noise. This signal carries the answer to a quantum computation, but the quantum computers are whispering the answer in a crowded room of chattering freeloaders. This means the signal falls exponentially fast as the number of qubits increases. However, for ten qubits, a signal should still be possible, and ten qubits is plenty to do some very interesting things. For example, a proof in principle demonstration of Shor's factoring algorithm using a six-qubit system can be done using a nine-nuclear spin system, with a relative signal strength of about 10 per cent.

Gershenfeld and Chuang show how to build Hadamard gates and CN gates for the molecular spins. Hadamard gates are easy as they only require rotating the magnetic orientation of individual qubits. The naturally occurring interactions between individual spins on the molecules can be used to build CN gates. The spin of one nuclei provides a magnetic field for its neighbour, which then tries to partially align itself with that magnetic field, and vice versa. The various quantum circuits required for quantum factoring can be implemented by carefully timed pulses of radio waves with different frequencies. The trick is to address one or two spins at a time. Fortunately, the nuclear spins in a molecule are not all created equal. Each finds itself in a slightly different chemical environment and thus a slightly different magnetic field. This means different external radio frequency field pulses can address particular nuclear spins. For a six-qubit factoring algorithm, Gershenfeld and Chuang estimate a few tens of pulses are needed to prepare the initial state, about 40 more are needed to perform the computation and a final few tens of pulses are needed to readout the result. This is well within the range of current NMR techniques.

The future

The three proposals to realise a quantum computer I have discussed are, of course, not the only ones around. At least two other schemes show long-term promise. One of these is based on a development in semiconductor technology known as quantum dots. The other is based on a super-conducting device called a SQUID (superconducting quantum interference device). Quantum dots operate near to the quantum limit, but not near enough. SQUID devices require a supply of liquid helium to keep them at their operating temperature of minus 269 degrees Celsius. Currently both ideas have a long way to go to overcome the decoherence problem. On the other hand, both are much more amenable to the essential mass miniaturisation required if a quantum processor is ever to be a practical device. The prospect is appealing enough that a number of groups are actively pursuing both technologies for possible quantum computation. In the end, none of the current schemes may be used for a practical quantum computer and the winning design will go to some as yet undiscovered scheme based on some undiscovered new material, for ultimately the decoherence problem is a materials problem.

EPILOGUE

Matter and energy can be organised to support information processing and computation, perhaps even thought. It is an idea so new, yet so obviously true, that few have grasped its significance, even as we rush to embrace the latest silicon marvel. We have carried the proof of this fact in our heads for millennia. In our brains, physical laws have been harnessed by evolution to spontaneously build a complex computational system, the principles of which are still poorly understood. Universal computers are not simply permitted by natural law, they may well be required. Perhaps late twentieth-century technology has simply got with the natural flow of the universe. The physical world however, is a quantum world. If computational systems are a natural consequence of physical law, then a quantum computer is not only possible, but inevitable. It may take decades, perhaps a century, but a commercially viable quantum computer is a certainty.

A quantum computer is not simply the next step in computers but a whole new paradigm for computation. Certainly a quantum computer could solve what a classical computer could never solve. But there is more to it than that. The idea of quantum computation is of deep significance for the foundations of physics. Feynman realised this, as does David Deutsch. Rolf Landauer, writing in 1985, put it very succinctly: 'not only does physics determine what computers can do, but what computers can do, in turn, will define the ultimate nature of physical laws'.[1]

192

Landauer is not talking about the prospect of a quantum hyper-computer solving very complicated physical problems, but rather the character of physical law itself. In Chapter 4 I raised this possibility in the story of the smart graduate and her office-bound adviser. This story was meant to illustrate the surprising consequences if the Church–Turing principle turns out to be true. This principle was hinted at by Feynman and stated explicitly by Deutsch. In a nutshell, it says that a perfect virtual reality simulator is possible in principle. The actual behaviour of the physical world can be reproduced by the behaviour of one specially organised part of it, a universal quantum computer, operating on finite elements of matter and processing finite amounts of energy. Computation, like measurement, is not a separate abstract ideal, but an integral part of the fabric of reality.

Yet it is abundantly clear that the laws of physics, as we currently have them, are not like this. Space and time are regarded as infinitely dividable and laws are formulated in terms of an infinite continuum. In such a universe the smallest subset of the space–time continuum already contains infinite information, which is forever beyond the capacity of a finite computational machine, quantum or otherwise. If the Church–Turing principle is true and Landauer's hunch is correct, we are about as far from a Theory of Everything as it is possible to be, infinitely far in fact. At this point I must trail off into speculation, but the stakes are high.

Seth Lloyd of MIT has followed the theme a bit further.[2] Lloyd suggests that in a universe in which local variables support universal computation, a quantum Theory of Everything can be simultaneously correct and fundamentally incomplete. This surprising result follows directly from the Halting theorem which I discussed in Chapter 4. The theorem says that it is not possible to know in advance if a given universal computer trying to solve a particular problem will finish and print the answer, or rather continue forever. The best you can do is run the calculation and watch the ensuing stream of apparently random binary numbers, with no possibility of grasping the pattern at the end because there is no end. Could it be that the ultimate physical theory is like that? Could it be that, even with the Theory of Everything, the best we could do is just watch the universe slowly reveal itself to us, surprise upon surprise, without end? What would such a universe look like at its most fundamental level? Would it forever produce a surprising answer to every question put to it in the form of measurement, a universe capable of producing an endless stream of random binary numbers? Intelligible, yet random . . . that sounds a lot like a quantum universe. I would be very happy to live in a world capable of endless surprise. Perhaps I do.

GLOSSARY

ASCI

Accelerated Strategic Computing Initiative. ASCI is a USA Department of Energy initiative which will create predictive simulation and virtual prototyping capabilities based on advanced weapon codes, and it will accelerate the development of high-performance computing far beyond what might be achieved in the absence of a focused initiative.

Bayes' rule

The probability for an event which can happen in two indistinguishable ways is the sum of the probability for each way considered separately.

beam-splitter

An optical device that splits an incoming light beam into a reflected beam and a transmitted beam of equal intensity without any reduction in the intensity overall.

Bell states

Four distinct quantum states corresponding to the four ways two physical systems, each of which can be in two mutually exclusive states, are correlated as strongly as possible. Classically there are only two ways to do this; either both systems can be in the same state or both systems can be in different states. However, as quantum states correspond to probability amplitudes, which can be

195

positive or negative, there is an effective doubling of the possible states.

binary code
A code system for representing information in terms of only two symbols, usually 1 and 0.

bit
A single binary digit of a binary code.

Brownian motion
Brownian motion refers to the erratic motion of small particles due to random collisions with the molecules of the surrounding medium. It was first observed by the biologist Robert Brown (1773–1858) and was explained by Albert Einstein, using the atomic hypothesis and statistical reasoning.

Caltech
California Institute of Technology, Pasadena, California.

cavity QED
The quantum electrodynamics of atoms in optical cavities. The optical cavity is a resonator for light formed by reflecting light between two or more mirrors. Quantum electrodynamics is the quantum theory of light and its interactions with matter.

Church–Turing principle
A physical principle proposed by David Deutsch: 'Every finitely realisable physical system can be perfectly simulated by a universal model computing machine operating by finite means'. The principle, if correct, implies that virtual reality is possible.

CN gate
A Controlled-NOT gate, also called an exclusive or gate, acts on two binary digits (bits). One bit is the control and the other is the target. The control bit is never changed, but if the bit on the control is 1 the bit on the target is inverted.

conditional gate
Any operation on two bits such that one bit is changed depending on the con-

	ditional value of the other, which is usually left unchanged, a controlled-NOT gate for example.
cyberspace	The world of information and communication enabled by using a computer as part of a network.
decoherence	Decoherence is the observed destruction *collapse* of superpositions of pure quantum states due to interactions with uncontrolled or unknown physical effects. Decoherence is responsible for apparent classical behaviour in an otherwise quantum world.
Deutsch gate	One of the first proposed conditional gates for quantum computation. The gate is universal for computation, which means that any computation can be done by a suitable network of such gates.
diffusion constant	A particle undergoing Brownian motion will on average move a distance from its starting point proportional to the square root of the time elapsed. The constant of proportionality squared is the diffusion constant.
ebit	An ebit is a single binary digit obtained as the result of a yes/no measurement on one part of an entangled quantum state of two systems.
EPR argument	The Einstein, Podolsky, Rosen argument refers to a paper by A. Einstein, N. Rosen and B. Podolsky, published in 1935, which purported to show, using entangled quantum states, that quantum mechanics was not a complete theory if physical states were given a classical interpretation. It was the first paper which brought the puzzling nature of

	quantum entanglement out into the open.
factoring	Finding the prime factors of large integers. This is regarded as a very difficult problem if the number has many digits. It is much easier to check that the factors are correct than it is to find the factors in the first place. This one-way property makes factoring an essential tool for cryptography.
Feynman's rule	The probability amplitude of an event that can occur in two or more indistinguishable ways is the sum of the probability amplitude for each way considered separately. Feynman's replaces Bayes' rule of classical probability theory.
Fredkin gate	A reversible logic gate which is universal for computation. It has three inputs and three outputs. One input is called the control. If the bit on the control is 1 the bits on the other two lines are swapped, otherwise the gate does nothing.
gate	A gate is a physical device used to implement a mathematical function on binary digits. The word is often used to refer to the mathematical function itself.
GHZ scheme	A proposal by Daniel Greenberger, Michael Horne and Anton Zeilinger to produce an entangled state of more than two qubits which exhibits correlations that cannot be explained by classical probability reasoning. The original proposal involved four qubits. Subsequently Mermin showed that it would also work for three qubits.
Hadamard gate	The simplest quantum logic gate for manipulating a single qubit. It simply produces an even or odd superposition

198

of the two mutually exclusive logical values of the qubit. It is often used together with a controlled-NOT gate to give a set of universal quantum gates.

halting problem   Given any Turing machine and any input to this machine, it is not possible to decide in advance if this calculation will stop or run forever. The issue is undecidable.

Hilbert space   In the quantum theory, every physical state of a system has an abstract mathematical description in terms of probability amplitudes for measurement outcomes. The set of all possible states form a mathematical system know as Hilbert space. The number of physical states available in quantum theory is vastly larger than the physical states postulated in classical Newtonian physics.

Landauer's principle   When information is erased energy will be given up as heat. There is a fundamental lower limit to how much energy will be lost; for each bit erased it turns out to be about the kinetic energy of an air molecule at room temperature.

Laplace's rule   Laplace's rule of insufficient reason states that, if we know no better, the probability for obtaining a particular outcome in a game of chance or other random process is the same for every outcome. All results are equally likely.

MIT   Massachusetts Institute of Technology

Monte-Carlo method   A numerical technique to solve mathematical and physical problems by simulating random process in the computer.

NMR   Nuclear Magnetic Resonance.

PBS   Polarising beam-splitter.

photon	Einstein explained the photoelectric effect by postulating that light transports energy in discrete identical packets called photons. The frequency (colour) of the light determines the energy of each photon. The intensity of the light determines the number of photons in the beam.
PKC	Public Key Cryptography. A system for encrypting and decoding data using a random key. The encryption key is available to everyone, but the decryption key is available only to the communicating parties.
plane polarisation	A beam of light is plane polarised if the electric force of the light is always aligned parallel to a fixed axis, perpendicular to the direction of propagation of the beam.
polarisation	In classical physics, light is a self-sustained wave of electric and magnetic forces. In a vacuum, the electric force is perpendicular to the direction of travel of the beam. The direction in which the electric force points is refered to as the polarisation of the wave.
prime factoring	It is very difficult to find the prime factors of a large integer, however, given a pair of prime numbers it is easy to check that they are factors. This one-way feature of prime factoring is used in a public key encryption scheme known as RSA.
probability amplitude	Quantum physics indicates that the Universe is irreducibly random. The theory enables us to calculate the probability of a given observation. However these probabilities are determined at a funda-

mental level by a probability amplitude, which need not be positive. The actual probability is found by squaring the amplitude which always yields a positive number.

quantum entanglement

If a physical system, composed of identifiable susbsystems, carries correlations between the subsystems, and these correlations can be realised in two or more ways, the state of the composite system is a superposition of the different ways to realise the correlation and the state is said to be entangled. Entanglement is the key feature of quantum physics that makes quantum computation so much more powerful than classical computation. The physical consequences of entanglement are still not fully understood.

quantum principle

The quantum principle states that the physical world is irreducibly random. No amount of knowledge of a physical state can make all possible measurements of that state certain. However the odds for measurement results are not given by the usual rules of probability but require a new probability calculus based on probability amplitudes.

qubit

A measurement which yields two mutually exclusive outcomes, with equal probability, requires one bit of information to store the result. In a quantum state however, the two mutually exclusive outcomes can be encoded as two probability amplitudes. In that case the quantum state encodes a qubit, which is experimentally distinguishable from a simple coin-toss.

Shor algorithm — The Shor algorithm enables the prime factors of large integers to be found in polynomial time. This is exponentially faster than the fastest known algorithms for a classical computer.

spin — A physical quantity, related to the rotational symmetry of a physical system. It is not predicted by classical Newtonian Physics, but is a new feature of the quantum world. In some cases the spin can only take on two mutually exclusive values.

superposition — If a physical state of a system can be realised in a number of different but unknown ways, then the actual state of the system is a superposition for each distinct way. There is a distinct probability amplitude for each way in which the physical state can be realised. The superposition principle is a fundamental feature of quantum theory and is a restatement of Feynman's rule.

teleportation — Teleportation refers to a form of communication between two parties that share an entangled state. It enables the state of a physical system in one location to be 'impressed' on another physical system at a different location through the transmission of one quantum entangled particle and one classical bit of information.

teraflop — A teraflop is a measure of computer performance. One teraflop corresponds to a million million (10^{12}) floating point operations per second.

Toffoli gate — A reversible logic gate which is universal for computation. It has three inputs and three outputs. Two inputs are called the control. If the bits on the control are both set to 1, the bit on the remaining input is inverted.

Turing machine An abstract set of transformation rules for manipulating strings of symbols to realise mathematical functions. The class of functions which can be obtained from a sufficiently complex set of such rules is called computable.

ultra-computer On 16 December 1996, Intel Corporation in collaboration with the Department of Energy (DOE) announced that the recently constructed Ultra-Computer was the first computer to reach 1 million million operations per second—one teraflop in technical terms. It is now housed in a laboratory in New Mexico, USA.

uncertainty principle A fundamental consequence of the way in which irreducible randomness enters quantum physics. The uncertainty principle implies that it is not possible to prepare a system in a state for which the results of all possible physical measurements are certain.

unitary dynamics Unitary dynamics describes the way in which quantum systems develop in time so that Feynman's rule will apply at all later times if it applies at an earlier time. Unitary dynamics requires that the system leave no record, in the outside world, of the states it passes through in time. Unitary dynamics is necessary for quantum universal computation.

ENDNOTES

Chapter 1

1 Richard P. Feynman, 'Simulating Physics with Computers', *International Journal of Theoretical Physics*, vol. 21, 1982, p. 467.

2 See my book *Quantum Technology*, Allen & Unwin, Sydney, 1996. (Published in America as *Schrödinger's Machines* by W.H. Freeman, New York, 1997.)

3 In fact many have, and a simple hidden-variable theory won't work. By the end of this book you will, I hope, appreciate why.

4 Richard P. Feynman, *QED: The Strange Theory of Light and Matter*, Princeton, New Jersey, 1985.

5 Richard P. Feynman, *The Character of Physical Law*, MIT Press, 1965, p. 129.

6 These probability arrows 'live' in an abstract world. They should not be confused with the directions of the real electric and magnetic forces for the light they describe.

7 This refers to an essay by John Wheeler, on the interpretation of quantum physics, called 'Information, physics, quantum: the search for links', in *Complexity, Entropy and the Physics of Information*, edited by W.H. Zurek, Addison-Wesley, Redwood City, 1990.

Chapter 2

1 A particularly good recent account may be found in *The*

Infamous Boundary by David Wick, Copernicus, New York, 1995.

2 N. Bohr, 'Discussions with Einstein on epistemological problems in atomic physics', in *On Atomic Physics and Human Knowledge, Volume II: Essays 1932–1957*, Ox Bow Press, Woodbridge, Conn., 1987.

3 The photon picture of light is discussed in more detail in my book *Quantum Technology*.

4 There is another explanation which accepts hidden variables, but hides them in 'multiple universes'. Feynman's rule is then explained by the way in which different realisations of the hidden variables interfere with each other between universes. This interpretation does not make clear what interaction is responsible for this interference between particles in different universes. Most physicists do not accept the many universes view as physically justifiable. For an opposing view see David Deutsch, *The Fabric of Reality*, Oxford University Press, Oxford, 1997.

Chapter 3

1 Private communication, 1997.

2 Daniel M. Greenberger, Michael A. Horne, Abner Shimony and Anton Zeilinger, *Am. J. Physics*, vol. 58, 1990, p. 1131. N. David Mermin, *Am. J. Physics*, vol. 58, 1990, p. 731.

3 C.H. Bennett, G. Brassard, C. Creapeau, R. Jozsa, A. Peres and W. Wooters, 'Teleporting an unknown quantum state via dual classical and Einstein–Podolsky–Rosen channels', *Phys. Rev. Letts*, vol. 70, 1993, p. 1895.

Chapter 4

1 This example is taken from page 46 of R. Penrose, *The Emperor's New Mind*, Oxford University Press, Oxford, 1989.

2 An example is given on page 71 of R. Penrose, *The Emperor's New Mind*.

3 David Deutsch, 'Quantum theory, the Church-Turing principle and the universal quantum computer', *Proceedings of the Royal Society of London*, vol. 400, 1985, p. 97.

4   See David Deutsch, *The Fabric of Reality*, Penguin Books, London, 1997.
5   Richard P. Feynman, 'Simulating Physics with Computers', *International Journal of Theoretical Physics*, vol. 21, 1982, p. 467.
6   Neils Bohr, 'Discussion with Einstein on Epistemological Problems in Modern Physics', from *Quantum Theory and Measurement* edited by J.A. Wheeler and W. Zurek, Princeton University Press, Princeton, New Jersey, 1983.

Chapter 5

1   This paragraph is drawn in part from my book *Quantum Technology*.
2   Strictly speaking, the photon is not completely specified by giving its direction of travel. We must also give its polarisation. However, I will make no use of polarisation in this simple example.
3   In fact we only need three numbers as the sum of the square of the amplitudes must sum to one.
4   David Deutsch, 'Quantum theory, the Church-Turing principle and the universal quantum computer', *Proceedings of the Royal Society of London*, vol. 400, 1985, p. 97.
5   David Deutsch, *Proceedings of the Royal Society of London*, vol. 425, 1989, pp. 73–91.
6   This paragraph is drawn in part from my book *Quantum Technology*.
7   If you would like the gory details, try 'Quantum computation and Shor's factoring algorithm', A. Ekert and R. Jozsa (eds), *Reviews of Modern Physics*, vol. 68, no. 3, pp. 733–53.
8   There is a problem if the period turns out not to divide the total number of states exactly. However, an adjustment can be made to take care of this case.

Epilogue

1   R. Landauer, 'Computation and Physics; Wheeler's Meaning Circuit', *Foundations of Physics*, vol. 16, p. 551, 1985.
2   S. Lloyd, 'Quantum-Mechanical Computers and Uncompatability', *Physical Review Letters*, vol. 71, p. 943, 1993.

INDEX

Printed in the United States
149960LV00003B/16/A